YANKEE TIGERS II

YANKEE TIGERS II

Civil War Field Correspondence from the Tiger Regiment of Ohio

Edited by Richard A. Baumgartner

BLUE ACORN PRESS
Huntington, West Virginia

BLUE ACORN PRESS
P.O. Box 2684
Huntington, W.Va. 25726

ISBN 1-885033-32-X

YANKEE TIGERS II
Civil War Field Correspondence from the Tiger Regiment of Ohio

Illustrated. Includes bibliographical references and index

History — American Civil War 1861-1865

Manufactured in the United States of America

Cover photograph:
Company C, 125th Ohio, at Camp Harker, Nashville, Tennessee, in June 1865.
[Mass. MOLLUS, USAMHI]

Frontispiece:
Capt. Anthony Vallendar *(standing center)* and men of Company H, 125th Ohio.
[Mass. MOLLUS, USAMHI]

CONTENTS

Acknowledgments / *7*

Abbreviations / *9*

Introduction / *11*

One
'Initiated into the mysteries of a soldier's life' / *39*

Two
'If you love your country, aim low — aim well' / *75*

Three
'Men who could make such a charge could storm Hell
and take the Devil by surprise' / *105*

Four
'One of our fiercest tiger yells had the desired effect' / *125*

Five
'Reading sermons in stones and sending
our prayers on bullet wings' / *139*

Six
'If the 125th can go no further, there is no use in trying' / *160*

Seven
'Fighting with cold steel and clubbed muskets' / *184*

Eight
'General Opdycke, I have the honor
of presenting to you this flag' / *201*

Nine
'Those true feelings that actuate the hearts of Americans' / *212*

Postscript / *224*

Appendix / *231*

Bibliography / *279*

Index / *283*

ACKNOWLEDGMENTS

Sincere gratitude is extended to a number of people whose assistance aided materially in this book's presentation.

Foremost among those to be thanked is Ohio native Larry M. Strayer, a longtime friend, research collaborator and Civil War photography historian. He critically scrutinized the Introduction and offered several helpful suggestions. Beyond providing valuable editorial input, he also contributed 27 of 54 wartime photographs illustrating the text, many of them previously unpublished.

Other photograph contributors include John Halliday, Benecia, California; Ken C. Turner, Ellwood City, Pennsylvania; Janis Pahnke and Ray Zielin, Chicago, Illinois; Richard W. Nee, Atlanta, Georgia; the Ross County Historical Society, Chillicothe, Ohio; the former Lotz House Museum, Franklin, Tennessee; and the U.S. Army Military History Institute at Carlisle Barracks, Pennsylvania.

Timothy R. Brookes of East Liverpool, Ohio, graciously made available correspondence written by a private in the 125th Ohio's Company E. The letters help document use of Henry repeating rifles in 1863 by four members of the regiment, which otherwise was equipped with the Springfield rifle-musket.

Lastly, more than a few words of friendly encouragement were offered by reference archivist John E. Haas, who greatly assisted access to General Emerson Opdycke's papers at the Ohio Historical Society library. Excerpts from Opdycke's wartime diaries appear in print here for the first time. Haas' audio-visual department colleagues were instrumental in obtaining scores of photocopies from bound newspaper volumes housed in the library's collections. Their efforts prove that a researcher's work is never accomplished alone.

ABBREVIATIONS

AGR	Adjutant General's Report.
CSR	Compiled Service Record.
Mass. MOLLUS	Massachusetts Commandery, Military Order of the Loyal Legion of the United States.
NARA	National Archives and Records Administration, Washington, D.C.
OHS	Ohio Historical Society, Columbus.
OR	*War of the Rebellion: A Compilation of the Official Records of the Union and Confederate Armies.*
RG	Record Group.
USAMHI	U.S. Army Military History Institute, Carlisle Barracks, Pennsylvania.

INTRODUCTION

T he written order Captain Emerson Opdycke held in his hand
at Murfreesboro, Tennessee, could not have been more welcome. Dated August 5, 1862, the document assigned Opdycke to
northeastern Ohio to recruit for his regiment, the 41st Ohio Volunteer Infantry.[1] It also meant he was going home.

The 32-year-old commander of Company A had been in the
field continuously since the previous November, enduring harsh
rigors of campaigning and fighting in Kentucky, Tennessee, Alabama and Mississippi. During the second day of battle at Shiloh
in April 1862, Opdycke, acting as the 41st Ohio's major, snatched
one of the regiment's flags when the color bearer was shot down.
Springing ahead, he shouted to his comrades to advance. "The
whole regiment followed grandly," Opdycke explained shortly afterward. "It seemed to me that a thousand bullets passed within
an inch of my head every moment; but on we went amidst that
storm of balls and cannister."[2]

Shiloh was a sobering introduction to combat for the regiment,
which lost 142 killed and wounded out of 373 officers and men engaged.[3] Opdycke himself was hit by two spent balls, as he related: "I received one shot in my right leg, and one in my right
arm, which drew a little blood but no harm was done."[4] Neither
wound kept him from remaining on duty. His conduct in the battle attracted immediate attention of his superiors as well as those
below him in the ranks. And it underscored a personal credo of
command he displayed throughout his Civil War career — "If an
officer *leads,* his men will be pretty sure to follow."[5]

Well before Shiloh, Opdycke's leadership qualities were being
honed. Few, however, could have foreseen his future military acu-
men in August 1861 when he left a moderately successful dry
goods business in Warren, Ohio, to assist raising an infantry com-

pany from Trumbull County.[6] Married less than five years and father to a three-year-old son, Opdycke possessed no previous military experience. It was a trait common to the vast majority of the war's early volunteers, North and South, who were elected or appointed officers. But with the arrival September 16 of newly appointed Colonel William B. Hazen at Camp Wood near Cleveland, Opdycke and his fellow recruits of the 41st Ohio were quickly transformed from citizens to soldiers.

Hazen, a Vermont native who grew up in Portage County, Ohio, proved to be a demanding taskmaster. An 1855 graduate of the U.S. Military Academy, he spent four years fighting Indians in Oregon and Texas (where he was severely wounded) before being appointed an assistant instructor of infantry tactics at West Point in February 1861. Given command of the 41st Ohio that fall, Hazen brought to the regiment a strongly imbued sense of firm discipline based upon exacting instruction and drill.[7] The regiment's historian recalled that the colonel first "seemed as insensible to fatigue as a threshing machine, and to think his men were like thirty-day clocks, wound up to run a month without stopping, and then wind up again. For lack of willing disposition, Hazen had no mercy ...."[8]

In addition to gaining thorough knowledge of army regulations and tactics, Hazen's officers were expected to study such military works as A.H. Jomini's *The Art of War,* William Napier's Napoleonic *History of the Peninsular War,* and special treatises on subjects including field fortifications, topographical drawing and minor engineering.[9] Mandatory classroom recitations were conducted regularly and, as Hazen phrased it, "this record became an unerring indication of the future of these officers." Those "not suited to their places, either from want of industry, character, or other causes, soon made their unfitness evident in many ways; and when there was no more hope, they either voluntarily resigned or were informed in a kind way that they were not likely to be useful, and quietly went home."[10]

An early "casualty" of the strict regimen was Company A's original captain, Seth A. Bushnell, who resigned November 27, 1861.[11] Opdycke, by contrast, flourished under Hazen's tutorial method. As the company's first lieutenant he wrote to his wife from Louisville, Kentucky, three days after Bushnell's exit: "... it seems to pay to attend faithfully to the duties for which we came. We are especially fortunate in our Colonel, he appreciates us for what we are, and what we do. You know we have regular classes for study and recitation &c. We have already gone through the 'Tactics,' Articles of War, and Dictionary of Military Science. The Class is composed of thirty-two officers, including the Lt. Col. and Major. The Col. keeps a regular record of the recitations and he showed me his Class Book the other day; and it is with no small gratification, that I tell you, that the name of 1st Lieut E. Opdycke stood first ...." [12]

The dedicated pupil, nearly 10 months older than his mentor, performed so well that in December 1861 and January 1862 he was detailed by Hazen to drill two Indiana regiments at their commanders' request.[13] At this time Hazen moved to brigade command and Opdycke was elevated, on January 9, to Company A's captaincy. His ambitious nature soon was aroused by thoughts of further promotion, which he displayed late in January while conversing at Camp Wickliffe, Kentucky, with his division commander, Brigadier General William "Bull" Nelson. Opdycke asked the general if he thought he was qualified to command a regiment. Nelson replied, "Yes, a damned sight better than these colonels here." Opdycke then inquired if Nelson "had any objections to saying it to the Governor of Ohio, whereupon he wrote me a very good recommendation." [14]

The captain continued to contemplate advancement following Shiloh. "How I would like to have a regiment of my own, drilling for the next great Contest," he wrote April 29.[15] This desire was whetted in May, June and July when Opdycke, in the absence of its field officers, temporarily commanded the 41st Ohio on several occasions in Mississippi and Tennessee. Late in July he applied for leave to return home to recruit for the regiment, whose ranks were much depleted since leaving Ohio.[16] He also was well aware that just three weeks earlier President Abraham Lincoln had called for 300,000 more volunteers to bolster manpower in the Federal armies. His native state was expected to furnish its fair quota, and these men would require leaders.

Believing himself fully competent to undertake whatever tasks lay ahead, Opdycke no doubt was overjoyed to be heading home once his leave was approved. Enroute he stopped briefly in Columbus and apparently did a bit of politicking at the Statehouse, visiting Governor David Tod. On August 12 he arrived in Warren and was reunited with his wife and young boy. Within 24 hours he opened a recruiting office in the storefront of friend and former business associate Servetus W. Park.[17]

One week later Warren's leading newspaper, the *Western Reserve Chronicle,* announced that "Authority was given on Friday [August 15] for the organization of fourteen new regiments to be recruited within seven days. The general order says, unless otherwise ordered, the surplus officers and enlisted men recruited for infantry regiments heretofore ordered, and now organizing, and the new unattached companies of volunteers raised, and to be raised, are assigned to the new regiments, from the 112th to 125th. The 125th regiment is to be formed in [the counties of] Trumbull, Ashtabula, Mahoning, Geauga and Lake."[18]

Although the announcement signified the opportunity Opdycke was hoping for, he persisted wooing volunteers for the 41st Ohio in Warren and Trumbull County's outlying townships. The *Chronicle* aided the effort with advertisements and periodic editorial notices, which extolled his previous service and urged his promotion. On August 27 the newspaper noted:

Capt. Emerson Opdycke, of the 41st Ohio regiment, has been at home for the last ten days, recruiting for his regiment. Notwithstanding the exposure, fatigue and privation of the campaign, in which the 41st bore so laborious, dangerous, and gallant a part, and some flesh wounds received at the battle of Shiloh, Capt. Opdycke has not been incapacitated from service for a single day, but has been engaged in active duty every day since the regiment went into the field. By his close study of military books, and careful use made of the efficient training of Col. Hazen, during the first few months of his military life, Capt. O. became so proficient that he was detailed by Col. Hazen to act as teacher in a military school which he established in his camp, and it was no unfrequent occurrence that lieutenants, captains, lieutenant colonels and colonels were under the tutelage of Capt. Opdycke.

By his conduct on the battle-field, he has shone himself to be a man of unflinching courage, and as by his actual services he has as fairly deserved promotion, as he is well qualified by nature and by study, to fill a higher position, we hope he may get one.[19]

Another item printed two weeks later informed readers that "Since Capt. Opdycke's arrival here ... he has recruited about 80 men for his regiment, the 41st. The noble young men of our county have responded to his call with cheering alacrity. During the past week he has had his recruits under drill, twice a day, on the public square, and the proficiency they have attained in that

ENLIST IN THE

GLORIOUS 41st!

———

HEADQUARTERS AT S. W. PARK'S HARDWARE STORE.

$92 paid on mustering in!

Choose your own company. I want 50 men in
Co. A.   EMERSON OPDYCKE,
Capt. Co. A. 41st O. V., Recruiting Officer:

41st Ohio recruiting advertisement published
in the *Western Reserve Chronicle,* September 10, 1862.

short time gives evidence that the reputation which had preceded
him, of being a drillmaster of the first class, was not unmerited."[20]

Echoing local consensus as well as positive opinions expressed
by several high-ranking army officers, the *Chronicle's* editor continued: "If more men are wanted, and the signs of the times give
evidence that they are, we hope Gov. Tod will issue orders for the
formation of the 125th regiment which was assigned to this district, and place Capt. Opdycke at its head .... We are credibly informed that Capt. O. has been recommended to Gov. Tod for promotion, by his Colonel, Brigadier and Major Generals, and yet
others, with but a moiety of his ability are filling, or rather occupying, much higher positions. Is it right for a subordinate to so
constantly discharge a superior officer's duties, and reap no honor
from it? We think not."[21] Neither did the governor, as Opdycke
jotted in his diary September 15: "Given the 125th Regt. to raise
and command."[22]

On the 17th, military committee representatives from seven
northeast Ohio counties assembled in Cleveland to apportion
quotas to be raised in each. Opdycke's home county of Trumbull
was required to furnish two companies.[23] The first to muster at
Camp Cleveland was drawn largely from Mahoning County, and
designated Company A. Its members had expected to join the
105th Ohio, but when that regiment secured a requisite number
of men for muster in August, the surplus was assigned to the
125th Ohio by executive order.[24]

16

Lotz House Museum

Emerson Opdycke (1830-1884) was considered by many contemporaries to be among the finest regimental and brigade commanders in the Army of the Cumberland.

The district committee anticipated the 125th's ten companies to be filled by October 1.[25] If Opdycke shared this expectation he was disappointed. By the fall of 1862 his area of the state had been well combed of eligible recruits. Those remaining were just as eagerly sought by other regiments then forming. Patriotic fervor had waned markedly since the war's early days, and was not boosted by newspaper accounts of recent bloody fighting in Virginia, Maryland, Mississippi and Kentucky.[26]

On October 10 Opdycke expressed a measure of frustration to a friend in the 41st: "I find the work of organizing a new regiment quite a difficult task. It is almost a month since my first company was mustered in, and none of the others are ready yet, though three captains have each enough men to insure success, while three others are less hopeful. I have only seven companies started, and do not know where the remaining three companies are to come from. The territory assigned to me as recruiting ground is limited in extent, and has furnished several regiments already."[27] Progress lagged. Not until November 1 was Company C mustered, followed by Company B more than two weeks later. Both were raised in Trumbull County.[28]

In the meantime, Opdycke divided his time between Warren and Camp Cleveland, while making one or two trips to the state capital. On October 7 he met with Governor Tod in Columbus, and was mustered out of the 41st Ohio to take official command of the 125th. In his diary he noted: "Was paid off from July 1st to Oct 6th $415.26 Feel flush."[29] His appointment to the rank of lieutenant colonel was dated October 1; the jump in grade to colonel did not occur until January 1, 1863.[30]

While commuting from Warren to Cleveland October 13, Opdycke stopped in the Portage County village of Windham to inspect and purchase a horse for $175. The seven-year-old "noble animal" was the first horse he ever owned.[31] "Barney," as Opdycke later named him, served as his primary mount for most of the next 21 months. Wounded in the mouth while carrying Opdycke up Missionary Ridge outside Chattanooga in November 1863, "Barney" was killed during a skirmish July 18, 1864 at Nancy Creek near Atlanta.[32]

As recruits slowly filtered into training camp to be "thoroughly drilled and disciplined" under Opdycke's watchful eye,[33] recommendations were made by the district military committee to fill the positions of his two subordinate field officers. The candidates chosen were Ezra B. Taylor for lieutenant colonel, and Captain George L. Wood for major. Like Opdycke, both were residents of Warren.[34]

Taylor, 39, was a prominent attorney who had practiced law in Ravenna, Ohio, for 17 years before moving to Trumbull County in January 1862. Although one member of Company A believed "no better selection could have been made from civil life," Taylor turned down the 125th's lieutenant colonelcy. In 1864 he enlisted as a private in the 171st Ohio, a 100-days' regiment composed of national guardsmen from the Western Reserve. Much of its time was spent guarding Confederate prisoners on Johnson's Island near Sandusky, Ohio.[35]

Wood, unlike fellow attorney Taylor, already owned a combat record. The 25-year-old Geauga County native was an early-war enlistee in the 7th Ohio Infantry, and as captain of Company D was shot through the thigh June 9, 1862 in the battle of Port Republic, Virginia. A severely injured sciatic nerve led to Wood's discharge November 10; however, before month's end he reentered the service and joined the 125th. Opdycke respected Wood, but his tenure as major was short-lived once the regiment took to the field. Lameness and constant leg pain forced him to resign his commission in April 1863. Wood returned to Ohio, briefly assisted with recruiting efforts and began writing a history of the 7th Ohio, which was published in 1865. Two years later he died at the age of 30, and was buried in Warren's Oakwood Cemetery.[36]

The positions of lieutenant colonel and adjutant were filled somewhat unexpectedly when, in December 1862, several groups of re-enlistees that served earlier in the 85th and 87th Ohio regiments (both three-month organizations) were ordered to Camp Cleveland to be merged into the 125th.[37] The 87th's detachment, mostly hailing from Knox and Richland counties, was headed by Colonel Henry B. Banning, a Mount Vernon lawyer.[38] He and his men had an interesting if inglorious story to tell.

Originally mustered early in June 1862, the regiment's 90-day service term had expired by mid-September when Lieutenant General T.J. "Stonewall" Jackson's Rebel corps besieged Harpers Ferry, Virginia. The 87th was among 12,500 Federal troops in the Harpers Ferry garrison that were surrendered to Jackson September 15 and paroled to await exchange. Once the regiment's circumstance was made known a week later, the officers and men were released from parole, sent home from Annapolis, Maryland,

George L. Wood, the 125th Ohio's first major, was photographed by F.L. LeRoy in Warren, Ohio.

and mustered out at Columbus.

Governor Tod immediately authorized Banning to raise a new 87th for three-year service, but recruiting progress was agonizingly slow. Because other regiments then forming faced the same plight, it was decided to consolidate Banning's detachment of 200 re-enlistees and recruits, and that of the 85th Ohio, with Opdycke's fledgling organization. "The 87th men have served a short term and they show it," observed a 125th Ohio private upon the veterans' arrival December 5 at Camp Cleveland. "Their clothes fit, their belts do not appear to chafe; when assigned to barracks, they broke ranks quietly and were at home, without excitement or confusion; their officers are good-looking young men, who seem to know their business."[39]

In the consolidation, negotiations were necessary to determine which officers would be retained. Banning ultimately accepted the 125th's lieutenant colonelcy, while Edward G. Whitesides, former first sergeant of Company I, 87th Ohio, was selected adjutant. He was a native Kentuckian residing in Pittsburgh, Pennsylvania, when the war began.[40] Nearly all of Banning's detachment and that from the old 85th merged to form Companies E and F, which mustered December 17. Companies G and H also mustered that month. They were formed by consolidating three understrength companies of recruits that had been in camp more than two months.[41]

By Christmas Eve, 751 enlisted men had been mustered, though only 630 were present for duty. Some were sick in the camp hospital, a few were transferred to cavalry regiments, and the rest deserted. Most were arrested in Cleveland.[42] "These men were given the name of 'bounty jumpers,'" recalled a Company B sergeant. "We had them in plenty to deal with."[43] Company A private Daniel K. Bush surmised the deserters were "hired" to enlist by committeemen to fill up ward or township quotas. "Perhaps it is just as well for the reputation of the regiment that they are gone."[44]

Opdycke spent six days at the close of the year with his family in Warren. "Christmas and at home!" he elatedly wrote in his diary. "It is so good to be here. There is a quiet joy in it that I love ...." The holiday reverie ended December 30 when he "Bade good bye to all at home and left for Camp. It is very sad but duty calls and all else must yield."[45]

The 125th's commander found marching orders awaiting him

the last day of 1862 at his Camp Cleveland headquarters.[46] Although the regiment contained only eight companies, they were to be ready to move at a moment's notice. Wishful assurances from the governor's office gave hope that two additional companies would be furnished soon after the other eight were in the field.[47] As it turned out, however, recruiting, organizing and mustering those two companies, I and K, consumed nearly a year. Supervision of the difficult, frustrating job fell to David H. Moore, an Athens, Ohio, minister who eventually became the 125th's lieutenant colonel.[48] His correspondence to Opdycke from June to November 1863 is found in the Appendix.

On January 3, 1863, the *Cleveland Herald* briefly alerted readers that "The 125th Regiment, Col. Opdycke, left for Cincinnati this morning. On arriving from Camp they marched around the Park and proceeded to the Depot, where they took the cars. The men looked finely and marched well, although the roads they passed through were deep in mud."[49]

Perhaps this was a fitting portent of things to come, for many muddy marches would be endured by the regiment over the next two and a half years. Its first destination in "Dixie" was Kentucky, where the 125th received its flags[50] and exchanged antiquated .69-caliber Belgian muskets for more accurate M1861 Springfield rifle-muskets.[51] Some, including Opdycke, hoped they were going to join Major General William S. Rosecrans' Army of the Cumberland, which had just fought a bloody three-day battle at Stones River near Murfreesboro, Tennessee.[52] The colonel learned his old regiment, the 41st Ohio, had performed creditably at Stones River, at the expense of 120 casualties.[53]

He also was well aware the new year brought a different objective to the Union war effort, which was hailed enthusiastically by the *Chronicle,* his hometown newspaper: "The great political event that shall signalize the first day of January, 1863, in American history, will consist of the Emancipation Proclamation of President Lincoln. It inaugurates a year of jubilee to millions of the down-trodden and oppressed sons and daughters of Africa .... But a day of deliverance has at last come, not as a measure of justice to them, but as an inexorable necessity .... The doom of Slavery has been written, and henceforward its career will be downward, deeper and deeper in an abyss from whose dark confines there will be no resurrection."[54]

Opdycke wholeheartedly supported the *Chronicle's* stance,

sharing his own views with his oldest brother, John: "I am anxious for peace, but on no other basis than the annihilation of rebellion and its black cause — Slavery. I am convinced that to have a permanent peace we must destroy slavery. And I don't want anything patched up for further generations to fight about. Let us give every human being liberty, and only [then] can we rely upon the intelligence and justice of the people. If we do this we will enter upon a long period of the most amazing prosperity and power."[55]

That still seemed a long way off as Opdycke embarked on a new phase of his military career with an untried regiment composed partly of veterans and partly of relatively raw recruits. Before many months elapsed, however, his pride soared exponentially as the regiment he molded consistently drew praise and approbation from superiors at brigade, division, corps and army level. Under Opdycke's guidance and dedicated adherence to drill, discipline and duty, his men displayed such tenacious ferocity and "heroic conduct" in their first major fight at Chickamauga that they were dubbed "Tigers" by their division commander. The nickname stuck like a badge of honor pinned to the uniform coat. A sobriquet emblematic of toughness, reliability and a remark made with determination in the heat of conflict — "They can kill us, but whip us, never!"[56] — it was carried into every camp and battle thereafter. One officer remembered that following Chickamauga, "the 125th seldom passed another command without hearing such expressions as 'There go the Tigers,' 'How are you, Tigers?' 'Go in, Tigers!' etc."[57] Three decades later a carved granite tiger was mounted atop the regiment's Chickamauga monument on Snodgrass Hill,[58] its mouth open in a perpetual growl, muscles flexed and gaze fiercely fixed south toward the Confederate foe.

During the war's final 18 months Opdycke was given more command responsibility, while the 125th Ohio added 15 battle honors to its banners and record, including Missionary Ridge, Dandridge, Rocky Face Ridge, Resaca (where Opdycke was again wounded), Peachtree Creek, siege of Atlanta, Jonesboro and Nashville.[59] On June 27, 1864 the regiment stormed uphill to attack Confederate earthworks at Kennesaw Mountain, Georgia, covering its brigade front as skirmishers. "Subjected to a tornado of cannister and minie balls," the Ohioans, like a majority of Major General William T. Sherman's Federal assault troops that day,

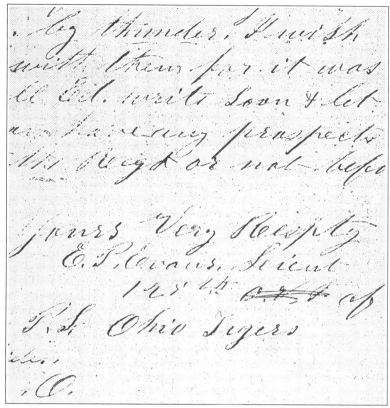

Written November 30, 1863, Lieutenant Ephraim P. Evans' letter to Adjutant
Edward G. Whitesides contains one of the first known references to the regiment's
sobriquet — Ohio Tigers — coined at Chickamauga by Brigadier General Thomas
J. Wood. "I pine for you and for our 'Ohio Tigers' like a caged bird," Lieutenant
Colonel David Moore wrote to Colonel Opdycke on November 27, just nine and
a half weeks following the battle. Such early references suggest that Wood's
originally expressed nickname was not "Opdycke Tigers," as he claimed 30 years
after the war.

were repulsed with heavy casualties.[60] In large measure their role
was reversed five months later at Franklin, Tennessee. At a crit-
ical juncture in the desperate battle fought there November 30,
Opdycke and the 125th Ohio covered themselves with glory by
charging headlong from a reserve position to plug a breach in the
Union line. "I ... never felt the effects of exertion in battle half as

much as on that occasion," Opdycke confided, but his regiment and brigade were credited by many with saving the day.[61]

Long after the battlefield smoke cleared — more than 30 years later, in fact — Charles T. Clark's exhaustive efforts of compiling and writing a regimental history were realized with publication of *Opdycke Tigers 125th O.V.I.* in 1895. Clark, who also belonged to the 85th Ohio,[62] spent most of his service under Opdycke's command as an officer in Companies F and H.[63] His book was considered "one of the best of Ohio regimental histories" by noted state Civil War bibliographer Daniel J. Ryan, who lauded its literary style, liberal use of photographic portraits and painstaking attention to detail.[64] For the general narrative Clark borrowed judiciously from the army *Official Records* (published between 1880 and 1901), and the military scholarship of Jacob Dolson Cox, a Warren, Ohio, attorney, Union general, Ohio governor and Department of the Interior secretary in the Grant Administration.[65] Aiding materially in recounting the regiment's story were wartime letters and diaries contributed to Clark by 11 former officers and enlisted men.[66]

Nearly a century passed before another book appeared in print with Opdycke's regiment as subject matter. This was *Yankee Tigers: Through the Civil War with the 125th Ohio,* a memoir written by Ralsa C. Rice, and edited by Richard A. Baumgartner and Larry M. Strayer (Blue Acorn Press, 1992). Rice, a sergeant, lieutenant and captain of Company B, originally composed his narrative late in life for 1905 inclusion in *The National Tribune Scrap Book,* an obscure, short-lived series of pamphlets issued under the auspices of the Grand Army of the Republic. Like Opdycke, he was a Trumbull County native and derived great satisfaction from his personal and regimental deeds. Rice wrote with remarkable accuracy, humor and candor, readily admitting his old regiment did not "put down the rebellion alone." The 125th Ohio, he modestly stated, "filled one short regimental front, but we kept that place warm."[67]

Emerson Opdycke certainly shared this opinion. His wartime correspondence to his wife, Lucy, published in 2003 under the title *To Battle for God and the Right,* convincingly illustrates his evolution as a leader, and those he commanded becoming competent, dependable, respected soldiers who rightfully earned the nickname "Tigers."

The body of information about the 125th Ohio contained in

these volumes is sizeable, but of course, not complete. A *definitive* account of the regiment, or any that participated in the Civil War for that matter, will never be fully told. But over time the emergence of new or long overlooked material incrementally enhances the story previously in print, thereby enlarging the published collective scope of written and visual documentation. *Yankee Tigers II* is intended to be such a supplement to the works of Clark, Rice and Opdycke — its sources primarily culled from the columns of several northeast Ohio newspapers which brought the 125th Ohio's bivouac, campaign and battle experiences home to readers with relative immediacy.

The *Cleveland Herald* and Warren's *Western Reserve Chronicle,* both pro-Lincoln Administration papers, were particularly welcomed and eagerly read by members of the regiment in the field. They regularly printed letters furnished by at least 10 different officers and enlisted men of the 125th. A majority of these regimental "reporters" veiled authorship by signing their correspondence with pseudonyms. Identities of the two most prolific writers can be established positively, while conjecture must suffice for others.

The *Herald's* first correspondent was Henry Glenville, a printer employed by the paper who used the pen name "Cato." Born Henry Glenville Shaw in Liverpool, England, he emigrated to America as a child and eventually settled in Cleveland. Three months shy of his 19th birthday he enlisted as a private in Company D, 84th Ohio, which performed 90-day duty during the summer of 1862 in the Department of West Virginia, guarding fords and bridges along the Potomac River's north fork. Upon muster-out in September he rejoined the *Herald's* production and editorial staffs, only to reenlist with the 125th on November 12. He was appointed first sergeant of Company H the same day.[68]

Glenville's earliest dispatches were subscribed with his initials "H.G.," but he soon adopted the moniker "Cato," likely borrowed from the name of Marcus Porcius Cato (234-149 B.C.), a Roman politician, soldier, writer and skilled orator.[69] His responsibilities as orderly sergeant aside, the diminutive Glenville (who stood less than five and a half feet tall) frequently was allowed time away from Company H to ply his civilian trade while the regiment was stationed outside Franklin, Tennessee, in the spring of 1863. On one occasion, this nearly cost him his life.

Confederate Major General Earl Van Dorn's mixed cavalry and

Lieutenant Henry Glenville, as he appeared near the end of his service in 1865.

infantry command assaulted Franklin April 10, portions of it reaching beyond the center of town before a Union counterattack dislodged the Rebels.[70] Colonel Opdycke related that Glenville "was over in Franklin printing some blanks for recitation reports for me. A rebel Captain took him prisoner, but his captor was soon shot dead and he escaped."[71] Clark, in *Opdycke Tigers,* recalled a slightly different version of the incident. The sergeant, "who was at the newspaper office ... doing duty as a printer, concluded to join the regiment forthwith. On the way he was observed by a rebel trooper, who put spurs to his horse and, gaining rapidly, raised his saber to cut down the straggler, who seemed to be just within his grasp, when a shot, fired by Malcolm Thompson, of Company H, killed the pursuer."[72]

"Cato's" correspondence was abruptly interrupted for 18 months following Chickamauga, where Glenville was shot through the side September 20, 1863 at the southern base of Snodgrass Hill. Left on the battlefield and captured, he was paroled nine days later at Crawfish Springs, then transported by wagon to Chattanooga and Stevenson, Alabama, for medical treatment.[73] The *Herald's* October 7 issue acknowledged on its front page: "By the list of wounded in the 125th Ohio, it will be seen that 1st Sergeant Henry Glenville, our correspondent 'Cato,' with that Regiment, and one of the volunteers from the Herald Office to the Army of the Union, was wounded at the battle of Chicamauga [*sic*]. We hope his wound is not serious, for Sergeant Glenville is too brave and likely a young man to be disabled, just now. May he soon be again with his regiment 'in fighting trim,' and wearing the shoulder straps he has well earned."[74]

Glenville underwent a year's convalescence at hospitals in Indiana, Ohio and Kentucky before returning to the 125th at Atlanta in September 1864 as a newly minted second lieutenant. He was further promoted to first lieutenant of Company B in April 1865, and mustered out with that company.[75] After the war Glenville moved west, residing the last 35 years of his life in San Francisco. There he found employment in the U.S. Customs Bureau, and for many years was engaged in newspaper work, at one time serving as editor and publisher of the *Pacific Veteran,* a periodical devoted chiefly to interests of the California National Guard. He died March 10, 1907 at the age of 63.[76]

"Cato's" colleague in the 125th who wrote extensively for the *Western Reserve Chronicle* above the signature "Ceylon" was

Ridgley Ceylon Powers, the descendant of a Mayflower pilgrim. Born Christmas Eve 1836 in the Trumbull County hamlet of Mecca, Powers was among the regiment's most educated men. He attended Western Reserve Seminary and the University of Michigan in the late 1850s, and graduated with honors in June 1862 from Union College, a liberal arts school in Schenectady, New York, then considered an academic rival to Harvard, Yale and Dartmouth. Recruited three months later at Mecca, the 25-year-old Powers listed his occupation as student. He immediately was appointed second lieutenant of Company C, the color company.[77]

Promoted to first lieutenant in the spring of 1863, Powers missed Chickamauga while on detached service in Ohio marshalling conscripts to the front. Timing of this hiatus was opportune, for he had been suffering since June with severe bouts of bronchitis and diarrhea. Powers returned to Chattanooga, fully recovered, just in time to witness the Army of the Cumberland's spectacular assault and capture of Missionary Ridge, November 25, 1863.[78]

Colonel Opdycke thought highly of "Ceylon," numbering him among his warmest friends and confidants. "I like Powers very much," he wrote early in 1864. "His father [Milo] is a Copperhead, but *he is not.*"[79] Opdycke urged his promotion, giving him the regiment's adjutantcy that March. In the midst of the Atlanta campaign he jumped to captain, and Opdycke, elevated to brigade command in August, selected Powers to be his acting assistant adjutant general.[80] At Franklin, the colonel reported Powers' "high judgment and courage" were instrumental to "the achievements of this momentous day." He also praised his staff officer's conduct December 16, 1864 in the battle of Nashville. Powers "rushed over the works with the troops and captured a rebel major with a number of privates."[81] When Opdycke unanimously was voted to receive the 125th's retired colors at Huntsville, Alabama, in March 1865, it was Powers who delivered the presentation speech.[82]

Exactly two years following his June 1865 muster-out, Powers was bestowed the brevet ranks of major and lieutenant colonel for "gallant and meritorious services" in the Atlanta campaign and the battles of Franklin and Nashville.[83] By then he had left Ohio for Mississippi. Settling in Noxubee County he acquired property, raised cotton and gained prominence in state politics.

In 1869 he was elected lieutenant governor, ironically serving

Clark, *Opdycke Tigers*   Clark, *Opdycke Tigers*

Ridgley Ceylon Powers   Hezekiah N. Steadman

under Governor James Lusk Alcorn, an ex-Confederate officer who once commanded a brigade of "inexperienced shot-gun militia" in western Kentucky. In office less than two years, Alcorn resigned to accept a U.S. Senate seat, and Powers succeeded him as Mississippi's 27th governor. During the balance of his term (1871-1874) the former captain "enjoyed the esteem and confidence of his fellow citizens, although he was a member of the unpopular Republican party, and his administration was marked by the suppression of lawlessness and return to economical government."[84] Powers lived another 38 years, mostly in Arizona. He died at Los Angeles November 11, 1912.

Regimental members corresponding periodically to the *Herald* or *Chronicle* in addition to "Ceylon" and "Cato" included several others who concealed their identities with noms de plume — "H," "W," "Hugo" and "Victor." The last two may have been the same individual using separate pseudonyms in homage to Victor Hugo, the renowned French poet, playwright and novelist whose most famous work, *Les misérables,* was published in 1862.[85] Opdycke believed "Victor" was Hezekiah Newton Steadman,[86] a Trumbull County native and school teacher who resided at Middlefield in Geauga County. He served in the 125th as a Company B corporal, regimental commissary sergeant, Company E first lieutenant and finally, captain of Company K. He also performed 10 months' duty as aide-de-camp on Opdycke's staff.[87] Internal clues in "Victor's" letters printed in the *Herald* and *The Jeffersonian Democrat* (published at the Geauga County seat of Chardon), strongly suggest Steadman's authorship. It is, perhaps, no coincidence that while Steadman was recruiting in Ohio between October 21, 1863 and March 1864, nothing by "Victor" appeared in either paper.

Reporters "H" and "W" have not been identified, though both probably belonged to Company B or C. With two exceptions in 1864, all letters written by "W" were printed in the *Chronicle* during 1863. A solitary missive by "H" ran on the *Chronicle's* front page in the wake of Federal success at Missionary Ridge.

Correspondence crossing *Chronicle* editor Comfort A. Adams' desk was penned by named soldiers in the 125th as well. In May 1864, newly promoted First Lieutenant Rollin D. Barnes of Company G contributed a wryly humorous letter about personal tribulations encountered during his return to the front following six months' recruiting service, most of that time sick. Barnes' gray hair implied his age. At 44, the Trumbull County barrelmaker from Bloomfield was the oldest officer commissioned in the regiment.[88]

Two decades younger, farmer Alson C. Dilley of Company C wrote to family and friends living at Howland, just east of Warren. At least three of his descriptive letters were loaned by them to the *Chronicle* for publication. Dilley, promoted second lieutenant from sergeant in March 1864, served during the summer of 1862 as a private in the 84th Ohio. That December a fellow Howlander characterized him as "one of those to whom the wishes and feelings of others are second only to an enlightened sense of duty .... He is strictly temperate in all things, firm and cool. No obstacle can long obstruct the way that he wills to tread."[89] Such attributes naturally attracted the attention of Opdycke, who could not "speak too highly of this gallant young officer."[90] Less than four weeks after his last letter appeared in the newspaper, Dilley was shot through the head and killed at Kennesaw Mountain.[91]

Even Opdycke, on occasion, submitted his thoughts to the *Chronicle*. Intended for public consumption, one letter exhorted women in Trumbull County to refrain from urging loved ones in the regiment to seek furloughs (see Page 64). It is clear Opdycke and his wife avidly read their hometown paper and the *Cleveland Herald*. Scores of wartime clippings from both periodicals were collected and saved by them.[92] By November 1863, Opdycke was certain that frequent newspaper coverage "pleases the men of the regiment, and their friends at home. The regiment has made itself famous, and I want the men to enjoy the great pleasure of having it known, and appreciated *at home*."[93]

Although the passage of time has dimmed present-day appreciation perceptibly, many incidents of the 125th Ohio's history and contribution to final Union victory are preserved in the compilation of correspondence presented here. In one of his last letters written in 1865, "Victor" eloquently summarized his fellow soldiers' symbolic epitaph: "Those men that had so often faced the bristling ranks of the foe, with hearts steeled to deeds of noble daring, wept as they thought of their slain comrades. But the stern, compressed lip showed that the fire of valor they had caught from their former Colonel still burned in their hearts, and that they deserve the name they had won, of 'The Tiger Regiment of Ohio.' "[94]

Richard A. Baumgartner
Huntington, West Virginia

Notes to Introduction

1. Special Order No. 126, Emerson Opdycke CSR, RG 94, NARA.

2. Glenn V. Longacre and John E. Haas, editors, *To Battle for God and the Right: The Civil War Letterbooks of Emerson Opdycke* (Urbana: University of Illinois Press, 2003), p. 28. This publication contains Opdycke biographical material and more than 300 wartime letters he wrote to his wife Lucy.

3. Robert L. Kimberly and Ephraim S. Holloway, *The Forty-First Ohio Veteran Volunteer Infantry in the War of the Rebellion. 1861-1865* (Cleveland: W.R. Smellie, 1897), p. 24. Reprinted in 1999 by Blue Acorn Press.

4. *To Battle for God and the Right,* p. 28.

5. *Ibid.*, p. 33.

6. *Ibid.*, p. xxiv–xxv. Born January 7, 1830 in Hubbard Township, Trumbull County, Opdycke moved with his parents and six siblings to Williams County, Ohio, in 1836. In 1847 he returned to Trumbull County, accepting an invitation to live in Warren with a married sister. In 1851 he opened a saddle, harness and trunk business there. Between 1854 and 1857 he lived and worked in California, then returned to Warren and married Lucy Wells Stevens on March 3, 1857. Their only child, Leonard Eckstein Opdycke, was born September 26, 1858.

7. William B. Hazen, *A Narrative of Military Service* (Huntington, W.Va.: Blue Acorn Press, 1993), p. vii–x, 3.

8. Kimberly and Holloway, p. 9.

9. Ibid., p. 9-10.

10. Hazen, p. 9, 8.

11. *Official Roster of the Soldiers of the State of Ohio in the War of the Rebellion, 1861-1866* (Akron, Cincinnati, Norwalk, 1886-1895), vol. IV, p. 170. Hereafter referred to as *Ohio Roster.*

12. *To Battle for God and the Right,* p. 8.

13. Hazen was born September 27, 1830. The Hoosier infantry regiments drilled by Opdycke were the 17th Indiana under Colonel Milo S. Hascall and 47th Indiana, Colonel James R. Slack.

14. *To Battle for God and the Right,* p. 12.

15. *Ibid.,* p. 33.

16. *Ibid.*, p. 43.

17. Emerson Opdycke diary, Emerson Opdycke Papers, MSS 554, Archives-Manuscript Division, OHS. Hereafter referred to as Opdycke diary.

18. *Western Reserve Chronicle,* August 20, 1862.

19. *Ibid.,* August 27, 1862.

20. *Ibid.,* September 10, 1862.

21. *Ibid.*

22. Opdycke diary.

23. *Western Reserve Chronicle,* September 17 and 24, 1862.

24. Charles T. Clark, *Opdycke Tigers 125th O.V.I.* (Columbus: Spahr & Glenn, 1895), p. 2; *Ohio Roster,* vol. VII, p. 570.

25. *Western Reserve Chronicle,* September 24, 1862.

26. Specifically, the 1862 battles of Second Bull Run, Virginia, August

29-30; Antietam, Maryland, September 17; Iuka, Mississippi, September 19; Corinth, Mississippi, October 3-4; and Perryville, Kentucky, October 8. Ohio regiments, some losing heavily, were involved in all five engagements.

27. Opdycke quoted in Clark, p. 5.

28. *Ohio Roster,* vol. VIII, p. 422, 425.

29. Opdycke diary.

30. Opdycke's appointments to lieutenant colonel and colonel can be found in Box 1, Folders 2 and 3, Emerson Opdycke Papers, OHS.

31. Opdycke diary.

32. *To Battle for God and the Right,* p. 76, 135, 199. The horse's original name was "Major," which Opdycke changed to "Barney" in the spring of 1863.

33. Clark, p. 5.

34. *Western Reserve Chronicle,* October 1, 1862.

35. Henry Howe, *Historical Collections of Ohio* (Norwalk, Ohio: The Laning Printing Co., 1896), vol. II, p. 677; Clark, p. 3; *Ohio Roster,* vol. IX, p. 431, 434.

36. George L. Wood CSR, RG 94, NARA; Wood to Opdycke, December 4, 1863, Emerson Opdycke Papers, OHS; Lawrence Wilson, *Itinerary of the Seventh Ohio Volunteer Infantry, 1861-1864* (New York: Neale Publishing Co., 1907), p. 170, 442. Wood was born at Chardon, Ohio, in 1837. His age is listed incorrectly as 37 in *Ohio Roster,* vol. II, p. 213.

37. Clark, p. 11-12.

38. Prior to being appointed colonel of the 87th Ohio, Banning served as captain of Company B, 4th Ohio Infantry. *Ohio Roster,* vol. II, p. 93; Roger D. Hunt and Jack R. Brown, *Brevet Brigadier Generals in Blue* (Gaithersburg, Md.: Olde Soldier Books, Inc., 1997), p. 30.

39. *Ohio Roster,* vol. VII, p. 1; Clark, p. 11-13; Whitelaw Reid, *Ohio in the War: Her Statesmen, Generals and Soldiers* (Cincinnati: The Robert Clarke Company, 1895), vol. II, p. 490; *The War of the Rebellion: A Compilation of the Official Records of the Union and Confederate Armies* (Washington: Government Printing Office, 1880-1901), vol. XIX, pt. 1, p. 549. Hereafter referred to as *OR.*

40. Clark, p. 12, 14; Edward G. Whitesides CSR, RG 94, NARA.

41. *Ohio Roster,* vol. VIII, p. 431, 434, 437, 440; Clark, p. 13.

42. Clark, p. 15; *To Battle for God and the Right,* p. 47.

43. Ralsa C. Rice, *Yankee Tigers: Through the Civil War with the 125th Ohio,* edited by Richard A. Baumgartner and Larry M. Strayer (Huntington, W.Va.: Blue Acorn Press, 1992), p. 22.

44. Clark, p. 16.

45. Opdycke diary.

46. *To Battle for God and the Right,* p. 46.

47. Clark, p. 14-15.

48. David H. Moore CSR, RG 94, NARA.

49. *Cleveland Herald,* January 3, 1863.

50. On January 19, 1863, the *Cleveland Herald* reported: "There are now on exhibition in Sargent's show window the National Color, the Camp Color and Markers, comprising part of a stand of regimental colors ordered by Col. Opdycke for his regiment. These colors are to be forwarded to the

regiment now in camp near Louisville, Ky. They were manufactured by G.W. Crowell & Co. of this city." George W. Crowell & Company was a sewing machine manufacturer and dealer, located in Cleveland's "Marble Block" at 211 Superior Street. Next door (at 213 Superior) was John W. Sergeant, manufacturer and dealer of mirrors, portrait and picture frames, engravings and lithographs. *Boyd's Cleveland Directory and Cuyahoga County Business Directory 1863-64* (Cleveland: Fairbanks, Benedict & Co., Printers, 1863).

51. While camped January 5-28, 1863 on the southern outskirts of Louisville, four 125th Ohio enlisted men received permission from Opdycke to privately purchase and equip themselves with Henry rifles, a .44-caliber, magazine-fed repeater that could fire 16 rounds in succession before reloading. Private George B. Vallandigham of Company E proudly wrote home that month: "The Reg'ts are armed with Springfield rifles. 4 of us are armed with 16 shooters, the most effective gun I ever saw. You can fire them faster than a revolver. I have one of them. They cost $45. The Col. has given us permission to carry them. I would not take 80 dollars for mine now & go without. The Gov't furnishes us with cartridges. Us 4 are equal to 40 men with them." [G.B. Vallandigham to his father, undated January 1863 letter courtesy of Timothy R. Brookes, East Liverpool, Ohio].

Columbus-born farmer George B. Vallandigham was a first-year student at Kenyon College, Gambier, Ohio, prior to enlistment at nearby Mount Vernon in October 1862. He was a nephew of Ohio politician Clement Laird Vallandigham, one of the country's most well known "Copperhead" critics of the Lincoln Administration.

52. The battle of Stones River was fought December 31, 1862 to January 2, 1863 between Rosecrans' Federal army and General Braxton Bragg's Army of Tennessee. It accounted for 12,906 Union and 11,739 Confederate casualties. During the night of January 3 Bragg's troops withdrew from the battlefield and Murfreesboro, marching to the vicinity of Tullahoma, Tennessee.

53. Kimberly and Holloway, p. 42.

54. *Western Reserve Chronicle,* January 7, 1863.

55. Emerson Opdycke to John Opdycke, June 23, 1863, Emerson Opdycke Papers, OHS.

56. *OR*, vol. XXX, pt. 1, p. 708.

57. Clark, p. 106, 126.

58. Ibid., p. 441-443.

59. *Official Army Register of the Volunteer Force of the United States Army* (Gaithersburg, Md.: Olde Soldier Books, Inc., 1987), vol. V, p. 222.

60. Clark, p. 278-279.

61. *To Battle for God and the Right,* p. 253, 250; *OR,* vol. XLV, pt. 1, p. 245.

62. At age 17, Clark enlisted May 26, 1862 in the 85th Ohio. He was appointed first sergeant of Company I and mustered out September 23, 1862. *Ohio Roster,* vol. VI, p. 656.

63. Commissioned a lieutenant in Company F, 125th Ohio, Clark commanded Company H at Chickamauga. In March 1864 he was detached for duty in Columbus, Ohio, serving there until April 1865 as, alternately, an

acting commissary of subsistence and commander of U.S. arsenal forces. He was promoted to captain to date from September 1, 1864. From June to September 1865 he held two brigade staff positions in the 2nd Division, 4th Corps/Central District of Texas. Charles T. Clark CSR, RG 94, NARA.

64. Daniel J. Ryan, *The Civil War Literature of Ohio* (n.p., 1911), p. 63-64.

65. Clark, p. i. The regimental historian primarily consulted Cox's books *Atlanta* and *The March to the Sea. Franklin and Nashville,* both published by Charles Scribner's Sons in 1882.

66. Clark, p. i.

67. Rice, p. 7, 8, 17.

68. *Ohio Roster,* vol. VI, p. 621, 628; vol. VIII, p. 440; Henry Glenville CSR, RG 94, NARA; *In Memoriam, Henry Glenville Shaw,* California Commandery, MOLLUS, Circular No. 14, March 15, 1907.

69. *Funk & Wagnalls New Encyclopedia* (Chicago: R.R. Donnelley & Sons, 1983), vol. 5, p. 374.

70. *OR,* vol. XXIII, pt. 1, p. 222-227.

71. *To Battle for God and the Right,* p. 68.

72. Clark, p. 53-54.

73. Ibid., p. 128; Henry Glenville CSR.

74. *Cleveland Herald,* October 7, 1863.

75. Henry Glenville CSR; *Ohio Roster,* vol. VIII, p. 422.

76. *In Memoriam, Henry Glenville Shaw.*

77. *The National Cyclopaedia of American Biography* (New York: James T. White & Company, 1906), vol. XIII, p. 493; Ridgley C. Powers CSR, RG 94, NARA.

78. Ridgley C. Powers CSR; *Western Reserve Chronicle,* January 6, 1864.

79. *To Battle for God and the Right,* p. 157.

80. Ridgley C. Powers CSR.

81. *OR,* vol. XLV, pt. 1, p. 241, 244.

82. *Cleveland Herald,* April 1, 1865. Within hours of the presentation, Powers left Huntsville for Ohio on 20-days' leave, taking with him at Opdycke's request the 125th's tattered flag, the colonel's sword, sash and other personal items. In Columbus he met with Ohio Adjutant General Benjamin R. Cowen, who "promised" Powers the colonelcy or lieutenant colonelcy of the 196th Ohio, then organizing at Camp Chase. Ten days later he instead was offered the majorship, but declined it in preference to remaining Opdycke's acting assistant adjutant general. Powers to Opdycke, March 18 and 29, 1865, Emerson Opdycke Papers, OHS; *To Battle for God and the Right,* p. 279.

83. Powers' brevets were awarded June 22, 1867 by General Order No. 65 of the War Department's adjutant general's office. Ridgley C. Powers CSR.

84. *OR,* vol. VII, p. 688-689, 813; *The National Cyclopaedia of American Biography,* vol. XIII, p. 493. James L. Alcorn (1816-1894) was born in Illinois and grew up in Kentucky. The founder of Mississippi's levee system, he served as a Mississippi militia brigadier general during the Civil War. Alcorn State University near Natchez is named for him. *Biographi-*

cal Directory of the United States Congress 1774-Present,
http://bioguide.congress.gov.
85. *Funk & Wagnalls New Encyclopedia,* vol. 13, p. 264-265.
86. *To Battle for God and the Right,* p. 114.
87. Hezekiah N. Steadman CSR, RG 94, NARA.
88. Rollin D. Barnes CSR, RG 94, NARA.
89. *Western Reserve Chronicle,* December 24, 1862.
90. *To Battle for God and the Right,* p. 189.
91. Alson C. Dilley CSR, RG 94, NARA; *Ohio Roster,* vol. VI, p. 627; vol. VIII, p. 425.
92. These newspaper clippings are found primarily in Box 2, Folder 9 and Box 4, Folder 1, Emerson Opdycke Papers, OHS.
93. *To Battle for God and the Right,* p. 126.
94. *Cleveland Herald,* April 1, 1865.

1863

ONE

'Initiated into the mysteries of a soldier's life'

Camp Opdycke,
Louisville, Ky., Jan. 8, '63.

Eds. Herald — On Monday morning last the 125th O.V.I. en-camped on the outskirts of this city. We left Cincinnati on Sunday morning, landing here the following morning.[1] The 124th O.V.I. is under marching orders, and leaves today for Elizabeth, Tenn. We are armed with Springfield muskets and shall probably follow in a few days.

Judging from the vicinity of this camp the late Major General Nelson had considerable ditching done round the city.[2]

The weather here has been warm and pleasant, but it is blowing a very cold wind to-day.

A great number of troops from Western Virginia are passing through here on their way to Murfreesboro. They are probably withdrawing their forces from that Department to reinforce Rosecrans.

H. G.

— *Cleveland Herald,* January 15, 1863

Camp Opdycke, Louisville, Ky.,
January 15, 1863.

Eds. Herald — The 125th O.V.I. are still encamped on the western outskirts of this city. It is very cold here to-day. Yesterday and the night before it has been raining felines and canines, and last night the wind changed and down came the fleecy element so that we are now blessed with a surface of eight inches snow and four inches mud. If this is a specimen of winter in Kentucky we shall be glad to move further south. Business appears to be pretty dull in the city just now, and everything and everybody looks miserable. If anything worthy of notice recurs in the track of the 125th you shall hear from

H. G.

— Cleveland Herald, January 19, 1863

On Board Transport Clara Poe,
Jan. 29th, 1863.

Dear Chronicle — Our stay at Louisville is finally at an end. We struck tents yesterday at 2 o'clock P.M., and at nightfall turned our backs upon Camp Opdycke, the place of our first encampment after leaving our own loved State. Our regiment had been on duty, during the day, acting as escort to the remains of the late Col. McGee,[3] and on account of fatigue, were in poor condition for moving, but the idea of getting into active service, away from the dull monotony of Camp, buoyed them up, and they took hold of the work with almost more than accustomed energy. But judge of our dismay, when on arriving at the boat-landing, we found ourselves ordered on the lower deck of the Jacob Strader, among about a hundred mules. All the inhabitable parts of the boat had already been occupied by the 121st and 98th Ohio regiments,[4] and we were left with the cheering consolation of knowing that the best we could do, was to accept the position assigned us.

Thanks to the humanity and energy of our Colonel and Major, at an early hour this morning we received the gratifying intelligence that better quarters had been procured for us and we were immediately set at work removing our baggage to the Clara Poe. Companies B and C (both from Trumbull) are in the Cabin, and the other companies are comfortably quartered about the boat.

By order of General Boyle,[5] the 125th has been assigned to the command of Brig. Gen. Gilbert, Tenth Division.[6]

Our present destination is Nashville, and all probabilities we will soon be engaged in the field.

We have all confidence in our Colonel, and Major, and we are proud to be led by such tried soldiers and able commanders before the enemy. We may not all return again to our homes, but I believe we will all try to be brave, and if we fall, it will be with a consciousness that our lives are spent in a just cause, worthy ten thousand times the sacrifice.

We leave in hospital, at Louisville, privates Asahel B. Hall, and Archibald Hill. They will probably be discharged.[7]

Respectfully Yours,
CEYLON

— *Western Reserve Chronicle,* February 4, 1863

On Board Transport Clara Poe,
Near Fort Donelson, Feb. 5, '63.

Eds. Herald — On the 23d January, the 125th regiment O.V.I. received orders to report to Brigadier General Gilbert, commanding 10th Division, when they assigned it to the 34th Brigade, commanded by Col. W.P. Reid,[8] of the 121st Ohio, and composed of the 98th, 121st and 125th Ohio.

On the 28th ultimo, we left Camp Opdycke, and after escorting the remains of Col. Samuel McKee, 3d Kentucky Infantry — killed at the battle of Murfreesboro — to burial, we proceeded to the wharf at Portland, three miles below Louisville, and embarked on board the steamer, thus forming a part of the reinforcements for Gen. Rosecrans' Army. On the 1st inst., the fleet left, and arrived at Smithland, at the mouth of the Cumberland River, on Monday evening. Here we awaited the arrival of other transports, and after coaling proceeded up the Cumberland on the following morning. Soon after passing the Tennessee line, and after dark, we perceived what we thought to be one of our boats on fire, but what afterwards proved to be a barge of hay saturated with turpentine floating down with the current. All the transports avoided it, and it did no damage. It was evidently the intention of the rebels that it should set some of our transports on fire, but it only served to brightly illuminate the surrounding country.

On Wednesday morning we were lying off Dover, half a mile above Fort Donelson, and at daylight descried, on the hills of that place, our skirmishers advancing as if anticipating an attack. We soon learned that a severe fight had taken place the afternoon and evening before, between the 83d Illinois and a Tennessee battery, our forces, and a superior number of Russell's rebel cavalry,[9] resulting in the repulse of the enemy with severe loss. I obtained permission to go ashore. The place was strewed with dead rebels and horses, some having met their death in an awful manner, being terribly mutilated. Here were thirteen lying in one heap, all having been killed by the explosion of one shell, and their dead were scattered everywhere round the place. The particulars of the fight, as near as I could learn, were as follows:

Col. Hardy,[10] [sic] commanding the 83d Illinois and Union forces in Dover, early on the day received intimation that twelve of his men were captured while guarding a Union family who were removing, and that a large force were advancing on the place. Col. Hardy made ample preparations for their reception with the small force at his disposal, and having judiciously placed his companies and guns in certain positions, awaited their arrival. About 2 o'clock P.M. they made their appearance on the south side, and forming in line of battle, sent a flag of truce demanding a surrender. This was at once refused; when having surrounded the place they opened fire from twelve pieces of artillery, and made several daring charges, but were each time gallantly repulsed by our men with heavy loss. At this crisis they sent in another flag of truce, notifying the Union commander that as yet they had not brought one-third of their force into action. To this the brave Colonel promptly replied that he had not used one-tenth of his, and that he would not surrender so long as he had a man left. The fight was resumed, but with no better success for them. Toward dusk they again demanded a surrender, and again received a decided negative. The advance of our gunboat escort here opened on them, shelling them out of the woods, causing them to retreat in splendid disorder, and ending the day with a triumphant *Union victory,* as if it were almost an anniversary of the victory of one year ago, which echoed through these hills and proclaimed throughout the North that Fort Donelson had fallen.

The 83d Ill., which is a new regiment having only been in the field since last August, deserves great praise for their gallant behavior on the occasion. In the fight they lost over twenty killed and wounded, including Capt. Reid of Co. A,[11] besides about twenty prisoners. The artillery men also handled their pieces with telling effect on the rebel ranks. The rebels must have numbered over 4,000 cavalry and artillery. This large force was kept at bay for over four hours, and ultimately badly defeated. Wheeler's cavalry is said to have comprised a part of their forces. Had they succeeded in capturing the place Fort Donelson would have been an easy prey. The line of fortifications commences at this point, and they might have greatly impeded our progress up the river and probably peppered us when we would least expect it, as we arrived only a few hours after the engagement.

L. M. Strayer Collection

Private George W. Chapman was 18 years old upon enlistment in October 1862 at Orrville, Wayne County. He served in Company F.

The rebel loss must have been immense considering the length of the conflict.[12] Up to this time, Thursday afternoon, our men have brought in over two hundred and twelve of their dead, and are still bringing them in. The gunboats must have done terrific execution on their retreat. On the retreat they treated their own dead shamefully and in a manner truly barbarous, throwing them in ditches, holes, and wells to prevent our discovering their loss. Over fifteen were discovered in one well and were dragged out by our men, and buried. Our cavalry are finding their dead and wounded for miles along their retreat. Reinforcements arrived from Fort Henry in the evening but too late to participate here, but they attracted the rebels' attention and skirmished with them on their retreat. It is reported here that the rebels afterwards attacked Fort Henry and succeeded in getting well whipped there also, but I cannot vouch for the truth of this.

Three rebel surgeons arrived here today under a flag of truce to attend to their wounded, over fifty of whom are in our possession. The 83d Illinois also captured 57 prisoners, including a captain and lieutenant, who are now under guard in Dover. They succeeded in capturing one field piece, horses, and equipments complete.

Among the remarkable incidents of the fight was that of a rebel cavalryman, who dashed up to the muzzle of one of our large siege guns, commanding our men to cease firing and surrender. He received a very forcible reply in the shape of the grape shot contents of his would-be prize, and there lay his mangled corpse and his horse, all shattered and torn a few feet from the gun.

This morning (Friday) the immense fleet is now all rendezvoused above the fort, and preparing to move, the smaller boats hitching together a la Siamese Twins, and we have a large escort of gunboats. If we are intercepted between this and Nashville, the rebels will have hard work.

H. G.

— *Cleveland Herald*, February 12, 1863

Steamboat Venando,[13]
Nashville, Tenn., Feb. 7, 1863.

Dear Chronicle: We have left our pleasant camp near Louisville and moved by way of the Ohio and Cumberland rivers to this place. The trip has been varied and interesting, occupying eleven days from the time we left Louisville. I may not be allowed to state the number of boats comprising our fleet, but anybody will admit that some seventy steamboats moving around a curve in the river, in full view, is a magnificent sight. We have had pickets established on the upper deck of our boats since we reached the mouth of the Cumberland, and have expected an attack; but in this we were disappointed, though we had a narrow escape at Fort Donelson, which was briefly as follows.

The rebels, knowing our destination, determined to make a desperate effort to stop and perhaps capture us. To accomplish this, they resolved to retake Fort Donelson, and gain possession of a huge siege gun which commanded the river. Accordingly a force estimated from 4000 to 6000 strong, consisting principally of Forrest's and Wheeler's cavalry, surrounded the Fort at about 12 M. the day preceding our arrival, and demanded its surrender.

The place was defended by the 83d Illinois, Colonel Harding, 550 men, and most bravely too, for from 2 o'clock until 8 P.M. did they withstand these tremendous odds; three times receiving, and as often repulsing them with great loss. But it remained for our gunboats to decide the day, and they arrived none too soon, for our little force was nearly exhausted, and could probably have held out but little longer ....

To one who visits a battle field for the first time, it is a sad yet interesting sight. But I will not attempt to describe what I there saw, and will only add that I have no desire to witness such scenes again.

We are now expecting to soon participate in one of the greatest battles of the war. The rebels will make a desperate effort to hold Tennessee, and Tullahoma is their last opportunity. May the right triumph, is our earnest prayer, and to accomplish this end will the 125th strive to perform its whole duty.

Yours truly,   W.

— *Western Reserve Chronicle,* February 18, 1863

Nashville, Tenn., February 7, 1863.

Eds. Herald — On the afternoon of Saturday last, many Nashvillians on casting a glance to the Northern sky, might have beheld a black, smoky cloud approaching them. A few hours after, if they had descended to their dirty wharves, a sight of splendor would dazzle their oculars. The large and magnificent fleet, composed of nearly all the finest river craft of the Ohio, loaded with

Union soldiers, had arrived from Louisville and were thronging the river opposite the city that afternoon.

At this city can be seen the wreck of progress and the marks of the willful destruction of the skill of man. On the opposite side of the river hangs the ruins of what must have been an ornament to the city at one time. There hang the rusty wires, bent and entangled, part in the water, of what once supported a handsome suspension bridge, and which now cannot fail to mock the city when the tale is told.

Our journey up the river was pleasant. Not even a random shot was fired at us, every precaution being taken and everything managed admirably.

H. G.

— *Cleveland Herald,* February 14, 1863

Camp near Nashville, Tenn., Feb. 10.

Dear Chronicle: We are at last arrived in the midst of Seceshia. We landed at Nashville on the 8th inst., having accomplished the trip without accident.

The country along the Cumberland is mostly very wild, presenting only occasional marks of cultivation. The rude log huts scattered along its banks remind one more of the wigwams of the aborigines than of habitations of civilized beings. The greetings we received by the inhabitants was not of that character calculated to awaken enthusiasm. Only on one occasion did we hear cheers for the Union.

We arrived at Fort Donelson on the evening of the fight, about five hours too late to participate in the victory. I was on the battle ground early next morning, and saw the ravages our gallant soldiers had made with the hordes of secession. Scores of dead rebels covered the ground; some with eyes out-torn, some with skulls opened to the wind, some pierced thro' the body, and others minus limbs, and mangled in every conceivable way. Horse and rider lay side by side, their blood comingling as they fell. On every side glassy eyes upturned to the sunlight only made the scene more horrid. Few countenances were placid. From most one would turn quickly away shuddering, lest the distorted image should become indelibly impressed, and haunt him as a spectre ever afterward.

Considering the forces engaged, it was a most glorious defeat of the rebels, and can scarcely be exaggerated. I hope it may be only one of a succession of disasters which the rebels shall experience, until they are forced to lay down their arms, and the Union shall be again established in all its former power.

I do not believe, as has sometimes been stated, that the Army of the Southwest is discouraged. From the best information I can gain, it is contented, willing to suffer more than it has yet suffered, that the country may be delivered from its present difficulties. The patriotic fire that burned so fervently two years ago has not abated; neither is love for the stars and stripes decreasing. We do not love our country less because it is afflicted, nor will we ever shrink from its support. Better that we should all perish than that a cause so nefarious, so unjust, so void of every principle of right as that urged against us, should succeed. Shame on those at the North who, having enjoyed the delights of a free government, are now unwilling to defend it from its destroyers; and, after a short struggle of two years, in which they have not engaged, are now willing to propose an Armistice or Peace Convention! Did they possess one drop of noble, patriotic blood, they would scorn to make such advances to a foe that had treated them only with insult.[14] Propositions for peace should not come from us; we are not the aggressors. To make such offers is but to bow at the feet of those who would trample us in the dust, and laugh at the cowardly spirit that had thus humbled itself.

Pardon this digression. I only intended to write a few facts more immediately connected with the 125th regiment.

We shall probably be joined to Rosecrans' command shortly; and, as the entire force with us does not fall much short of thirty thousand, you may expect to hear of work being done in this Department soon.

Respectfully yours,
CEYLON

— *Western Reserve Chronicle,* February 18, 1863

Franklin, Tenn., Feb. 14, 1863.

Eds. Herald — The Division to which we are attached took up the line of march from Nashville on the Franklin pike at daybreak on Thursday the 12th inst., our regiment occupying about the centre of the column. Nothing transpired to break the monotony of our military march until our arrival near this place, when

Company B's original organization contained 89 officers and enlisted men. Thirty-four of them hailed from the Trumbull County village of Kinsman, including Private John C. Mossman, who served until the company's June 1865 muster-out.

a report reached us that the town was occupied by some rebel cavalry. The 125th had the honor of being ordered forward as the advance guard. The regiments in our advance halted and stacked arms to allow us to pass. Among them we passed our neighbors of Camp Cleveland, the 124th Ohio. Upon our arrival on the outskirts of the town, our skirmishers (Company B) were fired on by the rebel cavalry pickets. The bridge across Harpeth Creek being destroyed, with Col. Opdycke at our head, we made a short cut across a field and waded the stream knee deep, a strong current running at the time. Our Colonel was as cool as on battalion drill, and while wading the stream gave the command "Platoons into line," the rebels at the same time firing at us behind stone walls, houses, &c., and skedaddling. Capt. Yeoman's[15] company (B) followed them sharply, and, after some brisk skirmishing, they retired on the Columbia road. There being no cavalry with the Division we were unable to follow and capture them; one of them was wounded, but he succeeded in getting off. Many of our muskets missed fire, as we marched from Nashville in a rain storm and our pieces were wet. A Brigade of Union troops, under command of Brig. Gen. Jeff. C. Davis,[16] had only left here in the morning, the rebel cavalry, about 300 strong, coming in half an hour afterwards.

The Confederates are in strong force at Columbia, about 20 miles from here. They may endeavor to cut us off from General Rosecrans, but if they have any regard for their personal safety let them keep a respectful distance.[17]

The 125th are now doing picket duty here, and the remainder of the Division are encamped on the other side of the creek. The rebel pickets are also near us and last night they made a dash on our pickets, fired and retired. Today a regiment of Union Cavalry arrived from Nashville,[18] and we now expect to make short work of their advances.

I have already taken up too much space, and shall forbear at present speaking of Franklin. It is rotten at the core; all the stores are closed, and the old white male citizens of the place are to be seen loafing at the corners of the streets and squares. No young men to be seen — all conscripted or volunteered in the rebel army. More anon.

H. G.

— *Cleveland Herald,* February 25, 1863

Franklin, Tenn., Feb. 18.

Dear Chronicle: Our company is on picket duty to-day, stationed on the Franklin and Columbia pike. A number of us, as a reserve, are seated around the camp fire, whiling away the time in talking, writing letters to friends at home, while not a few are listening to the reading of the *Chronicle,* which we received last night, and which, by the way, of all newspapers, is the most eagerly sought for by our boys.

Our regiment is fast becoming initiated into the mysteries of a soldier's life. Indeed, we are surprised at ourselves, when we think of the events of the past few days, and remember what we have endured since leaving our camp near Nashville. To be sure, an old soldier would smile, and say it is nothing, but to our friends will we submit the case, feeling sure they will render a different verdict.

We struck tents and moved out to the Franklin pike, according to orders received the day previous, and took up our line of march towards this place, a distance of 18 miles. It was our first march, and our knapsacks were well filled, which, with our other accoutrements, bore down heavily. After arriving within three miles of town, we received the intelligence that there were rebel cavalry ahead. The troops were halted, while our regiment filed past to the front, and Company B was immediately sent in advance as skirmishers. We moved on rapidly, and upon arriving in sight of town, took the "double quick" until we reached the bank of the river, when we deployed as skirmishers along the bank, and at the command of our Colonel, who at that moment came up, plunged into the stream, which was in many places waist deep, and with some little difficulty reached the opposite bank, where commenced our first skirmish in Dixie with the accursed traitors. We drove them through town and some three miles beyond, exchanging shots quite rapidly for a little while, but as they were mounted they had no difficulty in keeping at a pretty safe distance. As we had marched the greater part of the day in a drizzly rain, our guns were in a bad condition and did not work well, or we should have made a list of killed among the rebels. As it was, we wounded one only, while we came out without a scratch.

The rebel force consisted of Forrest's and Stearns' cavalry,[19] and had occupied this place but a few hours, though long enough to order a supper for all of their men; but as the Yankees came in sight about 3 P.M., and crossed so rapidly, they were obliged to leave the supper to the tender mercies of the "blue coats," who would have done it ample justice had they been made aware of the facts in the case at the proper time. The skirmishers returned to town at midnight, and took up quarters at the Female Seminary, where our regiment has since been stopping.

We are daily expecting an attack, and are fortifying as rapidly as possible. The people are secesh to a man, and we confiscate all that we desire. Twenty teams passed us to-day, returning from a foraging expedition, and loaded with corn, bacon, &c. This is the richest country that I ever saw, and if a fair specimen of Dixie, there is no danger of starvation at present, though there can be but little planted in this vicinity this spring for several reasons. Fences and negroes, both are gone.

We have 120 contrabands at work for us now, and still they come. But I have already written too much.

Yours truly,   W.

— *Western Reserve Chronicle,* March 4, 1863

[From the *Cleveland Herald,* March 11, 1863] — We are in receipt of the first number of the *Federal Knapsack,* a thoroughly Union paper, published in Franklin, Tennessee, by members of the 125th Ohio Regiment, Col. Opdycke. The advertisement of the paper is as follows:

The Federal Knapsack.
PUBLISHED AT
UNCERTAIN INTERVALS,
BY THE
125th Regiment, Ohio Volunteers,
AND DEVOTED TO THE
INTERESTS OF THE U.S. SERVICE.

Our Motto — "Up with our Banner and Down with Secession."

Terms — "No quarter to Traitors."

Exchanges — None but perfectly loyal papers will be entered on our Exchange List. Editors of the Cincinnati Enquirer school need make no application; every "true" soldier abhors them, as he does the miserable Bush-Whacker, who seeks by stealth to take his life.

The Editor is Lieut. R.C. Powers of Co. C, 125th Regiment, and the printing office is under the charge of Orderly Sergeant Henry

Glenville, a graduate of the Herald office. We copy from the *Knapsack,* which is dated March 4th, some paragraphs giving news in relation to the 125th Regiment and its present location:

Business in Franklin is entirely suspended. Aside from the military movements of the soldiers, the town presents, daily, the appearance of a continuous Sunday. It is the county seat of Williamson county, in this State, and eighteen miles south of Nashville. Its population before the Rebellion was said to have been fifteen hundred. All the citizens seem desirous for a speedy close of the war.

The construction of the railroad bridge across Harpeth river is progressing finely. The work is being done by three companies of the 1st Michigan Engineers, under the command of Capt. P.V. Fox.[20] The Captain is an energetic officer, supported by the right kind of men, as is apparent in the fact that he has been less than ten days in repairing the road from here to Nashville.

Permanent bridges, each of thirty feet span, were thrown across Little Harpeth and Spencer's Creek in the incredible short space of two days, the timber with which to construct them being taken from the stump and hauled some distance. In two days more the cars will be whistling through Franklin.

Our cavalry pickets had a lively time Monday last skirmishing with the enemy. Over two hundred shots were fired, killing and wounding none of our men. Two of the rebels were wounded and two taken prisoners. From one of the prisoners we learn that their attention was directed to some of our men on an eminence, north of the position they occupied, when some of our cavalry came suddenly and unexpectedly upon them, and took them prisoners before they had time to retreat.

Private Wm. Strahl, of Co. "E," 125th Regiment, exposed himself to their view, to draw them from behind the trees, where they had shielded themselves, for the purpose of experimenting with the "Henry Repeating Rifle," of which he and some others in the Regiment are in possession,[21] and which was the cause of their capture. The Rebel Cavalry — Texas Rangers [22] — after maneuvering in different directions, evidently for the purpose of ascertaining our position and strength, retired, pursued some distance by our cavalry.

There is a Brigade of Rebel Cavalry at Spring Hill, some twelve miles south of this place. They do not seem to relish our forces being in possession of Franklin. If they should have the hardihood to make a dash, for the purpose of taking it, they may look for a warm and bloody reception.

The rebels are reported to be in strong force at Columbia, twenty-three miles south of here, on the south bank of Duck river. It contains a population of about 2000. It is noted as the residence of James K. Polk, deceased, formerly President of the U.S. It is said to be a place of learning, wealth and fashion.

Capt. C.C. Baugh, of Company E, 125th Regiment, has been appointed Provost Marshal of this city. He is a man of business capacity and energy, and will make a prompt and efficient officer.[23]

Col. Opdycke is now in command of the Post at Franklin, and Lieut. Col. Banning is in charge of the 125th Regiment.[24]

[From the *Cleveland Herald,* March 21, 1863] — The second number of the *Federal Knapsack,* published at Franklin, Tenn., by the 125th Regiment, has been received. We again clip freely from its columns for the information of the friends of that regiment in this vicinity. The date of the paper is March 13. Of "the position" it says:

Although we are not at liberty to state anything definite about the number and situation of the Union forces in and about Franklin, we may be allowed to state the following facts, interesting to our friends, without injury to the service. The men of the 125th Regiment are comfortably quartered in good houses in the town. We have plenty of clothing, provisions, &c., with which to make soldiering, as far as possible, agreeable. Col. Opdycke, though for the time taken from us and put in command of the post, does not forget to look to our interests. Under his tuition, in the short space of five months, we have become one of the best drilled regiments in the service; and although we have done little fighting, there has been manifested a laudable zeal which is a sufficient guarantee that we will do our duty before the foe.

Gen. Gilbert, by assigning to us the post of honor on entering this place, and by repeated acts of favor since, has shown himself to have confidence in us and our gallant leader.

Lieutenant Col. Banning, during the short time he has been with us, has been making rapid advances in gaining the entire respect of the men. He is now in command of the regiment, and is indefatigable in his efforts to make all under him comfortable.

We miss the genial temper and manly heart of Major Wood. We lament the cause that separates him from us, and hope he may soon be able to return.[25]

We are, indeed, in a country surrounded by the foe, but our hearts are in the work which we have set out to accomplish, and let what will come, we will patiently and cheerfully bear all for our country's sake.

• Four men of Co. E, 125th regiment, were struck by lightning at half past nine o'clock on Saturday evening last. They were on picket duty, and stationed on the Columbia road, in a grove of small ash trees, a half mile from town. The lightning shivered the tree under which they sat, scattering the splinters in all directions for some distance round. They are all from Ohio, and their names and residences are as follows: William Nickerson, Mount Vernon; J.F. Randolph, Morrow county; R. Beeman, Mount Vernon; E.H. Dillen, Gambier. The last named was killed, and the first seriously injured, though hopes are entertained of his recovery.

The other two have so far recovered that no serious consequences are apprehended concerning them. The first named was a corporal, and the other three privates. Private D. was a young man of moral and exemplary character, a good soldier, and his sudden death cast a melancholy gloom over the entire company of which he was so worthy a member.[26]

• The manner in which Colonel Opdycke is dealing with the secessionists about Franklin is giving entire satisfaction to Union people, who, previous to our possessing the town, were imprisoned and paraded through the streets at the point of the bayonet. Even ladies were subjected to this most disgraceful process by the Southern *chivalry,* for no other crime than expressing a love for the Old Flag.

• Henry H. Thomas, sutler of the 125th Regiment, O.V.I., while taking some guns from a wagon on Friday last, accidentally discharged one of them with fatal result; the ball entering his right temple, inflicting a terrible wound, from which it is thought he cannot recover.

• S.N. Jones, Serg't. Co. "C," 125th Regiment, has been placed in charge of the "contrabands" at this Post. He has now on his roll, able for duty, with pick-axe and shovel, over 250 names.[27]

• Dr. McHenry, of the 125th Ohio, has been appointed Post Surgeon, and is now in charge of the General Hospital (formerly a College building) in this place.[28]

Private Daniel Cooper, Company F. In October 1863 the Mount Vernon farmer was detailed for 23 months' duty as regimental commissary assistant. Carte de visite by Lilienthal Gallery, New Orleans.

• Heman R. Harmon, Lieut. Company C, 125th Ohio, has been appointed Ordnance Officer for the 34th Brigade, 10th Division.

Camp near Franklin, Tenn.,
March 16th 1863.

Dear Chronicle: Have just been reading a copy of the Warren *Constitution,*[30] and I feel so indignant that good old Trumbull is polluted by the publication of so vile a sheet, a sheet so utterly void of sympathy for the true patriot, and our most holy cause, that I cannot resist the impulse to take my pen and through the medium of your paper, say to our friends at home, that they may say to our enemies in their midst, that we can but regard the publication of the *Constitution* and its supporters as truly our enemies as those who with the deadly rifle seek our life on the battle field. Yes more, and were it our privilege to choose between the two, which should be the mark for *our* rifle, we would not hesitate to aim first at those who seek to discourage the soldier by misrepresentations and falsehoods and who also strive to create dissension and strife at home, for we should feel that in so doing we were rendering our country the greater service.

The paper asserts that the soldiers are tired of war and are anxious for peace and compromise, and to make the assertion good, copies from the Cincinnati *Enquirer,* and other newspapers of like stamp, numerous letters from soldiers, in which they complain bitterly of the war, and of the sufferings they have to endure. Now we will admit that there are soldiers who are anxious for peace on almost any terms, among whom may be classed those whom the government *compelled* to enter the service, and also those who for a sum of money took the name of substitute. There are some of these classes — a sufficient number perhaps to furnish these papers with "letters from soldiers," but the great mass of Union soldiers are as firm in their purpose to stand by the old flag and fight so long as there is an armed rebel in the land as they ever were.

When our forefathers were struggling with a mighty power for the foundation of this Republic, there were as there is now, a class of men who cried peace, peace; but how much more base are those who having tasted the fruits of the best government on earth and lived under its protection for so many years, now seek (indirectly though it may be) its destruction and overthrow. But they will fail of their object. The soldiers in the field have already marked them, and they cannot much longer poison the minds of fainthearted men of the North. Such disloyal sheets must and shall be suppressed. *The soldiers have said it,* and let those men beware how they misrepresent our thoughts and feelings, and let them remember that as three million men were sufficient to gain our nationality, twenty million are able to preserve it.

Our regiment moved from Franklin yesterday one mile north, and across the Harpeth river. Are again living in tents which is far more pleasant, as the weather is quite warm now. Peach trees are in full bloom reminding us of the month of May. It is talked pretty strongly by some that Colonel Opdycke is to command a brigade. He is capable, and we hope to see him fill that position. He is highly esteemed by his men, and we are proud to be under such a leader.

On Saturday last the troops in and about Franklin were reviewed by Major General Granger.[31] It was a fine display. Today some 400 men are at work on the fort which will be a formidable affair.[32]

The health of the regiment is improving very fast; there has been considerable sickness, and we regret to announce the deaths of John C. Nailer, Co. A, and Wm. A. Covert, Co. B, both good and faithful soldiers.[33] In regard to our pursuit of Van Dorn,[34] and other rebel generals, you have probably already heard through other sources, and I will only add that we are all safe, and ready to meet the enemy at any hour.

Yours,   W.

— *Western Reserve Chronicle,* April 1, 1863

Camp near Franklin, Tenn.,
April 5th, 1863.

Eds. Herald — With the exception of sending out occasional scouts and foraging trains, the Union forces at this place have made no move since our visit to Spring Hill three weeks ago. All our operations have been confined within the narrow limits of Camp. Target practice, drill and working on fortifications comprise the sum total of our labors. So insensible have we been to what has been going on around us, as to allow Forrest with his rebel horde to make a raid nine miles to our rear and capture

three hundred men who were guarding a railroad bridge at Brentwood. [35] And this too, when a very moderate amount of caution might have prevented it. The truth of the matter is, our Generals are too sleepy; they have not got the bold dash and wide-awake courage necessary to conduct an army successfully in such a war as this. We have got too many West Point men, who, adhering strictly to scientific rules, are continually laying plans, dreaming of regular approaches, strongholds taken, sieges ably conducted, fortifications reduced and armies surrendered. The long parallels and offensive display at Corinth and Yorktown are illustrious examples of the folly of these educated military geniuses. Their *education,* of course, is not to be objected to, but the bungling ineffectual manner in which they apply it should justly expose them to ridicule.

We want Generals who will capture, not drive the enemy. Every time we drive them we weaken ourselves and strengthen them. Their lines of communication are shortened and their ability to mass large armies at any given point with celerity increased, while our forces are decreased as we occupy new places, guard bridges and protect our extended lines of communication and supplies. Rosecrans only seems to have a just conception of the character of the war; and above any other General in the West has devised measures, which properly supported, would send rebellion howling to its dark home. His order sending all the rebels, men, women and children out of our lines, will do more toward conquering a peace than half a dozen battles like Shiloh and Stone River. The establishing of battalions of mounted infantry[36] is another enterprise worthy the attention of the Government, and will have the effect most effectually to prevent all such raids as Kentucky has been exposed to from the commencement of the war.

Since the 125th Regiment left, the post at Franklin has been broken up. Our forces are now encamped on this side of the river (Harpeth) about one mile from the town. Our picket lines extend beyond Franklin, and citizens in sympathy with the rebels are not allowed to pass them. General Granger declares he will make the

Opposite: A 20-year-old farmer from Williamsfield in Ashtabula County, Private William A. Covert, Company B, died at Franklin March 3, 1863, of congestive fever and diarrhea.

place too hot to contain them long; he is determined to starve out all secessionists in the vicinity of his army.

The 125th now composes a part of Benison's Brigade, Col. 78th Illinois Vol., now acting Brigadier.[37] Brig. Generals Baird [38] and Gilbert are still in command of Divisions at this place.

Spring has opened beautifully, weather fine and roads good. All that is wanting is the word of command, and the army of the South West is ready to move.

Yours truly, CEYLON

— *Cleveland Herald,* April 13, 1863

Camp near Franklin, Tenn.,
April 7th, 1863.

Dear Chronicle: After a silence of some weeks, I again send you a few lines informing you of the condition of our regiment. Since leaving the Post at Franklin, we have been encamped this side of Harpeth river, nearly one mile northeast of town. The site is a beautiful one, commanding a fine view of the surrounding country. Pure springs gushing from lime-stone rocks in the immediate vicinity of the camp, furnish us with plenty of excellent water, and clear, babbling brooks running on either side of us offer splendid facilities for bathing. Spring has opened delightfully; the fields are putting on a covering of verdure, and opening buds in the forests and fragrant flowers in the valleys presage the glory of the coming season. With this kindly influence pervading the face of nature, there is a corresponding cheerfulness manifest in the hearts of the soldier-boys of the 125th. Yet there is not among them that same expression of gaiety they possessed while in Camp Cleveland. They do not sing as joyously nor talk as thoughtlessly as they did then. A change has come over the spirit of their dreams, and, becoming more earnest and manly, they seem to realize the danger with which they are surrounded, and the weight of responsibility that rests upon them.

It is with feelings of deepest grief that I have to inform you of the deaths of three of the members of Co. C — Jonathan Dilley, Joseph Andrews, and David Jack, the two former with fever, and the other with pneumonia. They all died within the space of one week, after a short but severe illness.[39] They were all young, the last named only being married, and their generous dispositions and attention to duty endeared them to both officers and men.

The health of the regiment is improving, there being only eleven men sick in hospital and thirteen excused from duty on account of slight indisposition.

In my next I will inform you concerning military movements in this department.

Respectfully Yours,
CEYLON

— *Western Reserve Chronicle,* April 22, 1863

Fort Granger, near Franklin, Tenn.,
April 26, 1863.

Eds. Herald — Put that down on the map. The pick and the shovel have done it. If Middle Tennessee is to be saturated with blood during the forthcoming promised campaign, the Fort which I have introduced to you at the heading of this letter is destined to tell a tale in the history of the rebellion.

This part of the country has not hitherto been the scene of any hostile engagement between the friends and enemies of the Republic. When Nashville fell into our hands the enemy did not select the Tennessee & Alabama railroad as their line of retreat, but retired on the Nashville & Chattanooga road to a place of safety. The people here are only commencing to realize the bitterness of civil war. General Buell's army carried the national banners in triumph through here to Columbia, but no traces were left that an army of conquerors had already gone before. The inhabitants of the country are warm admirers of that General, and miss no opportunity to draw unfavorable comparisons between us and the forces of our predecessor. It was not until the middle of Feb. that our forces entered into permanent possession of this place. On the 12th of that month Gen. Gilbert's Brigade [*sic*] was sent out here, and after repairing the railroad and rebuilding a bridge across Harpeth river, immediately commenced the work of fortifying, the nature of the country offering superior advantages for such operations.

Fort Granger is on the bank of the river, which, at that point, runs east and west, and is not approachable from that side. Here it commands the town and the open country beyond. If you would approach the Fort from any other direction, you will have to ascend to it at an angle of 45 degrees for some distance. The digging has been mainly done by our soldiers. When the 125th Ohio occupied the town, they daily confiscated niggers under the Proclamation and escorted them to the trenches, but the African citizens soon became scarce, and are now at a discount for defensive purposes.

Fort Granger, although yet in its infancy, has already greeted the rebels in language too *forcible* to be mistaken. On Van Dorn's recent visit, the brave General whose worthy name it bears acted as Corporal, and with his own hand sighted the siegers and tugged the fuse.[40]

As the length of my letter will not permit me to speak of any other subject but Fort Granger, I will conclude by stating that to Capt. Merrill,[41] of the U.S. Topographical Engineers, is to be attributed the credit of the engineering beauties which are daily becoming more apparent. The policy which seems to be just now adopted by our Generals here is fortify! fortify!! fortify!!!

CATO.

— *Cleveland Herald,* May 2, 1863

Camp near Franklin, Tenn.,
April 28, 1863.

Dear Chronicle: Four things have conspired to make April thus far very agreeable to the men of this command.

Pleasant weather and good health are two of them, for which we cannot be too thankful, although, usually, we are apt to consider them least. Third, we have been paid off; and lastly, we have been victorious over our enemies. What could our hearts wish more?

On the 10th inst., Van Dorn made a dash upon us with the intention, as we have every reason to believe, of capturing our entire force. But woe to the evil spirit that directed him, his plans were frustrated, his well appointed forces were put to confusion, and ere a single point he had hoped to gain was realized, the vaunted courage of his braves had left them, and they were fleeing like scared dogs before the determined bravery of what, previous to the war, they were pleased to call *"Cowardly Yankees."* His main force was met and put to flight before it had advanced to within four miles of our fortifications, while the few skirmishers that dashed boldly up to within range of our heaviest guns, were many of them killed and others captured. The few, who under the mad excitement of liquor, rushed up under the very shadow of our forts, sent back no messenger to convey to their comrades the story of their misfortune. The affair had the effect very much to dispirit the enemy in front of us, and, although Van Dorn undoubtedly has an army superior in numbers to ours, he will feel his way with more caution when he next makes up his mind to attack us.

One of Uncle Sam's paymasters made a very acceptable visit to this command lately. The 125th received pay on the 14th inst. up to the 28th of February. As it was the first given us since entering the service (over six months), it caused the face of many of our brave boys to lighten up as he received into his hand the shining green back — a just recompense for his trial. That we did not forget our friends, you will readily admit when I tell you that the two companies from Old Trumbull sent home over nine thousand dollars. Pretty good for the first installment, don't you think? But Oh! when we return ourselves bringing the happy tidings that our cause is victorious, our liberties preserved, and our country saved, what heart will not gladden, what eye not moisten at the warm greeting? The entire debt will then be paid. Copperheads and peace croakers will have no portion there.

A brilliant exploit conducted by Col. Watkins, of the 6th Ky. cavalry[42] terminated successfully yesterday morning. With a detachment of cavalry from this place he made a bold dash into a small camp established by Van Dorn about eight miles out on the Carter's Creek road. He found the rebs at roll call and had them surrounded before they were aware of his presence. He captured one hundred and twenty prisoners including nine commissioned officers, over one hundred horses and mules, besides saddles and other equipage. Tents, wagons and such baggage as he did not wish to bring away were burned. The expedition returned safely to camp about noon yesterday without having lost a man.[43]

Van Dorn still has his headquarters in the neighborhood of Thompson's Station, eight miles out on the Columbia pike. A very large cavalry force has just been sent to reconnoiter his position. The army never was in better spirits than now.

I am truly yours,   CEYLON

— *Western Reserve Chronicle,* May 13, 1863

[From the *Western Reserve Chronicle,* May 13, 1863] — The officers in the Army of the Cumberland, and we suppose throughout all of the armies, are constantly receiving letters from persons at home, requesting them to give furloughs to soldiers in their respective commands. From an eloquent letter written by Colonel

Opdycke to a lady in this county, who had requested a furlough for her husband, we are permitted to make the following extracts touching the matter:

Head Q'rs 125th O.V.I.,
Franklin, Tenn., April 29th '63.

Dear Madam: Your letter of the 22d inst. is at hand and contents noted ....

It is never in the power of a Colonel to give a furlough. General Rosecrans alone has that authority here. There are thousands of men who have families at home, and who have not had a furlough for *two years.* I expect a higher degree of patriotism from you ladies at home than to be over-persuading your relatives to get furloughs. The cause we are fighting for is of too great importance to be thus lightly treated. I want to see the ladies at home urge their husbands, sons and brothers to stand at their posts *till every rebel has ceased his treason and infamy.* This is worthy the American women. Let them see that their fame remains untarnished. Your sacrifices in staying at home are equal to ours who are in the field. Bear them cheerfully, nobly, and heaven will be your reward.

The nation's life is worth every life in it, and every dollar made under the folds of its protection is nothing when compared with the value of our nationality. If the nation is to go down amid the billows of contending ambition and passion, then will the free of the world be enveloped in the gloom of ruin. Should it be divided, each division will be driven to the protection of immense standing armies and these will drag us down, down to monarchies, to the horrid overthrow of the rights of man, as surely as the needle to its pole. Such results must and will be avoided, and the only question for us is, shall the contest be long or short? This will be in proportion as we act unitedly and wisely. The women of the South are keeping up the rebellion, those of the North *can* put it down. Your influence may send every man to the field. Exert it patriotically, and we will all be home next Spring. Tease us to come home and the war will last ten years. So if you want us to come home, *urge us to stay here.*

Very respectfully,
Your obd't serv't,
EMERSON OPDYCKE,
Col. 125th O.V.I.

Company D private C. Lafayette Gilbert spent much of his service after October 1863 detached as a nurse and musician in Chattanooga. In 1885 the Portage County farmer presented the 125th O.V.I. Association with a painted "Tiger" banner that was displayed at all successive regimental reunions. Carte de visite by McPherson & Oliver, New Orleans.

Gen. Granger's Army Corps,
Franklin, Tenn., May 19, 1863.

Eds. Herald — Nothing calling for particular attention has transpired in this command since the late Van Dorn's attack last month.[44] All is refreshingly quiet. Our pickets go and come from their duties undisturbed, and all symptoms of a fight must be *pro tem* laid on the shelf.

The weather could not be more delightful and favorable for active operations of an offensive nature than at the present time. All the roads are in splendid condition, and have been so for the last few weeks; the sky is clear, the rivers and creeks are low and the health and spirits of the army excellent. The troops are anxious to strike *another* blow in Tennessee, and one which shall suffice to leave a mark which secession shall carry with it to its grave.

In the Chief of the Army of the Cumberland we have every confidence, and his representative on this the right wing, Gen. Gordon Granger, enjoys that unlimited confidence which a General should possess in order to be successful.

I have spoken of pleasant weather, but that does not include all our comforts. Our inner man is well provided for. Rations are wholesome and plentiful. An occasional mess of good vegetables or potatoes vary our usual diet. Good wheaten bread is issued to us fresh daily. We are not lacking good and comfortable clothing. Our sick are well cared for. In short, no army could be better cared for than ours is at the present time. Railway communication with Nashville is undisturbed. Dan McCook[45] protects our rear. Mail facilities perfect. An abundant supply of newspapers can always be found in camp, including your welcome paper, so that in reading matter we are not behind the times. The agents of the Christian Mission[46] are laboring zealously in the cause in which they are engaged, much good resulting from their efforts to improve the religious tone of the army.

Sending superfluous baggage to the rear is a game which two can play at. We are packing such extra baggage to Nashville, which will be safely stowed at that point. If making formidable preparations to take the defensive, and practicing the "double quick" step, can be taken as an indication of an advance, then we are certainly preparing.   CATO.

— *Cleveland Herald,* May 25, 1863

Near Franklin, Tenn.,
May 23, 1863.

Eds. Herald — There has been a dearth of news in this Department for the past month. In the absence of anything more interesting I have concluded to give you a brief description of the
surroundings of "Camp near Franklin," earnestly hoping that I
shall not be anticipated "By Telegraph." Those "lying wires" are
really a great draw-back to the army correspondent. If they do
make many mis-statements for him to correct, they also sap the
life from everything like news, leaving him to supply only minor
matters of detail. But I am bound to climb above the level of the
profession.

To this end I have to-day, under a scorching sun, ascended and
taken position on Roper's Knob, a beautiful eminence four hundred feet above the surrounding country, and offering from its
summit one of the most delightful views in the West. To the
north, south and east, the valley — a very garden, now scarred by
the blight of war, dotted here and there with noble residences,
once the seats of luxury and wealth — stretches away, gradually
rising as it recedes, until it blends with the clouds beyond the cir-
cle of distinct vision. Toward the west the country is more broken, descending by a range of uneven hills toward the Tennessee
river. At the base of the Knob on the south side are the camps of
Smith's Cavalry Brigade,[47] and adjoining him are Baird's and
Gilbert's Divisions of Infantry.

From a high bank of the Harpeth, one-half mile south, Fort
Granger frowns defiantly, commanding with its terrible "siegers"
the country for miles. Five other fortifications of less extent on
neighboring elevations also present a formidable array, and with
the one at this point, complete the artificial defence of the place.
The Knob is our strongest position, and when the plan now in
process of execution is completed it will be the Gibraltar of Tennessee. One thousand men may here defy ten times their number, or resist an attack of even more in perfect security.

Roper's Knob is named in honor of a former owner — and
"hereby hangs a tale." Mr. Roper built his house on a small level
plat about two-thirds the way up the hill. Here for several years
he lived a happy life, surrounding himself with a family of bloom-
ing daughters, whose winning ways, in connection with the natural romance of the spot where he dwelt, attracted about him

scores of admiring visitors. But "a change came o'er the spirit of his dream." He lost his fondness for society, and with it his wonted cheerfulness. All was not satisfactory in his family. Possibly he was no longer monarch of all he surveyed. However this may be, his troubles were consummated when he awoke one morning to find his wife suspended by the neck from the limb of a tree near to his house, *stone dead.*

Thus was the name Rop(h)er's Knob made doubly appropriate. What has since become of the remnant of the unfortunate family, I do not know. The house has been demolished, there being left only an old chimney stack and a few ornamental trees to mark the spot where once it stood.

The U.S. Signal Corps have a post on the Knob, and by means of flags and telescopes they are able to give and receive information readily at great distances. At night they use lanterns. In time of an engagement these signal officers are of very great use in sending dispatches from one part of the field to another. Through them we are able to know all that is transpiring along our lines between here and Murfreesboro, and Rosecrans can be attacked at no point without knowing it almost immediately.

It is rumored that there is to be a general movement in this department very soon. May it be onward to victory is the earnest hope of your humble servant.

CEYLON,
125th Ohio Vol's.

— *Cleveland Herald,* May 27, 1863

Editors Chronicle: I send the following extract from a letter written by Sergeant A.C. Dilley, to his friends, thinking it may interest some of your readers.

Franklin, Tenn., May 24th, 1863.

Dear Friends at Home: Seated under a large tree on a hillside a half mile from home, or camp, my mind floats back on pleasant

Opposite: Captain Edward P. Bates, Company C, was often in temporary command of the 125th, leading it in the battles of Missionary Ridge and Franklin. Highly esteemed by Colonel Opdycke, the Hartford, Ohio, native received his captaincy in September 1862 from Ohio Adjutant General Charles W. Hill. During the summer of 1861 Bates served as a private in Company D, 19th Ohio Infantry (three-months' organization).

remembrances to the dear old hillside shades of my northern home. I am overlooking the most beautiful country I have ever seen. The limpid waters that gush from every hill afford such sweet draughts as remind me of the oval spring at the foot of the hill, where childhood's lips gathered in freshness; but we can claim nothing akin to this most wealthy and gentle climate. The placidity, yet luxuriance and variety of this climate entitles it to be termed feminine, if one may take liberty thus to compliment. Health seems to have "her seat and center in the breast" of Tennessee's hills ....

The health of our company is good. We prosper well under our good Captain Bates.[48] Lieut. Powers is now in command of Co. A. He is now a first Lieutenant, we have no second Lieutenant. We drill every morning from four to five o'clock in battalion. Our regiment forms in "line of battle," in five minutes from bugle sound. We do not feel inferior to any regiment in drill, or otherwise, but we leave others to sound our praise. We have work to do under gallant officers, and with hearty will we will do it. We are cheered with good news from Vicksburg to-day and hope it will soon be ours.[49] When that is "done taken," we hope to move I say not upon whom.

Your fond Son and Brother,
A.C. Dilley,
Co. C, 125th O.V.I.

N. B. Of letters and magazines, send us plentiful showers and of newspapers let there be no withholding. We devour news most eagerly. Do not fail to send us *Chronicles*. It is like shaking personified Trumbull Co. by the hand to read one.

A.C.D.

— *Western Reserve Chronicle,* June 10, 1863

Notes to Chapter One

1. Opdycke's January 5 diary entry noted: "At Louisville at 1 A.M. Camped a little way out by 3 P.M. Seems like old times to get in tents."
2. Major General William "Bull" Nelson, Army of Kentucky commander in late August-September 1862, was organizing Louisville's defenses against a threat of Confederate attack when he was fatally shot September 29, 1862 in Louisville's Galt House hotel after an altercation with Brigadier General Jefferson C. Davis. Mark M. Boatner III, *The Civil War Dictionary* (New York: David McKay Company, Inc., 1959), p. 586.
3. Lieutenant Colonel Samuel McKee commanded the 3rd Kentucky Infantry (U.S.) when he was killed in the battle of Stones River, December 31, 1862. Promoted colonel earlier that month, McKee had not mustered at the higher rank prior to his death. *Report of the Adjutant General of the State of Kentucky 1861-1866* (John H. Harney, 1866), vol. I, p. 592-593.
4. Until June 8, 1863, the 125th Ohio was brigaded with the 98th, 113th, 121st and 124th Ohio Infantry regiments.
5. Brigadier General Jeremiah T. Boyle, then commander of the District of Western Kentucky. Frederick H. Dyer, *A Compendium of the War of the Rebellion* (Dayton: Press of Morningside Bookshop, 1979), vol. I, p. 527.
6. Brigadier General Charles C. Gilbert, a native of Zanesville, Ohio. Ezra J. Warner, *Generals in Blue: Lives of the Union Commanders* (Baton Rouge: Louisiana State University Press, 1993), p. 173.
7. Private Asahel B. Hall, Company C, was discharged for disability at Louisville, February 10, 1863. Private Archibald Hill, Company C, was promoted to corporal and mustered out September 25, 1865. *Ohio Roster,* vol. VIII, p. 426, 427.
8. Colonel William P. Reid resigned November 4, 1863. *Ohio Roster,* vol. VIII, p. 275. After reporting to Reid January 28, Opdycke thought "He is not military," underlining the opinion in his diary.
9. Confederate troops assaulting Dover February 3, 1863 were commanded by Major General Joseph Wheeler, and Brigadier Generals Nathan Bedford Forrest and John A. Wharton. *OR,* vol. XXIII, pt. 1, p. 39. Union forces consisted of the 83rd Illinois (nine companies) and Battery C, 2nd Illinois Light Artillery. *Report of the Adjutant General of the State of llinois* (Springfield: Phillips Bros., 1901), vol. V, p. 149.
10. Colonel Abner C. Harding was promoted to brigadier general in May 1863 to rank from March 13. Warner, p. 207; *Illinois AGR,* vol. V, p. 125.
11. Captain Philo E. Reed, a former resident of Warren, Ohio, was killed February 3. He was an "old friend" of Opdycke who, before leaving Dover, "Saw his remains in its burial case." *Illinois AGR,* vol. V, p. 126; Opdycke diary.
12. Colonel Harding reported Union losses at Dover as 13 killed, 51 wounded and 20 captured. He estimated Confederate casualties at 150 killed, 600 wounded and 105 captured, while General Wheeler only admitted "about 100" killed and wounded. *OR,* vol. XXIII, pt. 1, p. 38, 41.
13. With the steamboats *Jacob Strader* and *Clara Poe* overcrowded, Opdycke on January 31 "sent Major [Wood] & 2 comp[anies] on the *Venango*

to make more room." Opdycke diary.

14. "Ceylon" referred to the North's growing number of Peace Demo-crats, derisively labeled Copperheads, that opposed President Lincoln's war policy and favored a negotiated peace with the Confederacy. Copperhead strength lay mainly in Ohio, Indiana and Illinois. Boatner, p. 175.

15. Captain Albert Yeomans was from Kinsman, Trumbull County.

16. At the time, Brigadier General Jefferson C. Davis of Indiana com-manded the 1st Division, 20th Corps.

17. The bulk of Major General William S. Rosecrans' Army of the Cum-berland was encamped in the vicinity of Murfreesboro, 26 miles southeast of Franklin.

18. The 9th Pennsylvania Cavalry.

19. Colonel James W. Starnes, 4th Tennessee Cavalry (C.S.).

20. Captain Perrin V. Fox, 1st Michigan Engineers and Mechanics.

21. See Page 34, note 51.

22. The 8th Texas Cavalry, better known as Terry's Texas Rangers.

23. Captain Calton C. Baugh, 40, was from Mount Vernon, Knox County. In May 1863 an examining board convened at Franklin found Baugh "not proficient in the Company drill & in the several duties of a Company Commander." Poor health may have been a mitigating circum-stance. That July Opdycke wrote that Baugh "has been reported for duty with his regt but 39 days in the last six months. He is entirely worthless as an officer, and has been recommended to resign by his superiors, on up to Department Head Quarters. He refused to do so, and is in Nashville. I therefore request that he be mustered out of the service of the United States." Baugh was dismissed from the army July 31, 1863. Calton C. Baugh CSR, RG 94, NARA.

24. Banning arrived at Franklin February 18, having been on detached service in Ohio up to that time. He left the 125th Ohio April 6, 1863 to take command of the 121st Ohio. Henry B. Banning CSR, RG 94, NARA.

25. Major Wood's thigh wound suffered at Port Republic, Virginia, the previous year forced him to leave the 125th Ohio February 11, 1863. He resigned April 20, 1863. George L. Wood CSR.

26. The lightning incident occurred March 9. Private Eli H. Dillon eventually was interred in the Franklin section, Grave 51, Stones River National Cemetery. *Ohio Roster,* vol. VIII, p. 770. William V. Nickerson, a Huron County native, was appointed sergeant April 1, 1864. Morrow County farmer Joseph F. Randolph was shot in the hip September 20, 1863 at Chickamauga, and was transferred in March 1864 to the 43rd Company, 2nd Battalion, Veteran Reserve Corps. Richard Beeman, a farmer born in Wayne County, was captured at Chickamauga and died of dysentery May 1, 1864 in prison at Andersonville, Georgia. W.V. Nickerson, J.F. Randolph and R. Beeman CSR, RG 94, NARA.

27. Sergeant Silas N. Jones transferred to the Veteran Reserve Corps April 15, 1864. *Ohio Roster,* vol. VIII, p. 425.

28. Opdycke considered Surgeon Henry McHenry to be a "treasure." He "keeps almost every man in the regt. fit for a fight." *To Battle for God and the Right*, p. 55.

29. First Lieutenant Hemon R. Harmon resigned April 16, 1863 due to

disability. *Ohio Roster,* vol. VIII, p. 425.

30. The *Constitution,* Warren's Democratic newspaper, frequently criticized the Federal government's war effort and occasionally provided editorial space for letters written by disgruntled or disenchanted soldiers serving at the front.

31. At the time, Major General Gordon Granger commanded the Army of Kentucky. Later in 1863 he commanded the Army of the Cumberland's Reserve Corps and 4th Corps.

32. Dubbed Fort Granger, the large earthwork was located along the Harpeth River's north bank just northeast of Franklin.

33. Private John C. Naylor died at Franklin February 21. Private William A. Covert, a farmer from southeast Ashtabula County, died March 3. Both are buried in Stones River National Cemetery. *Ohio Roster,* vol. VIII, p. 422, 423; William A. Covert CSR, RG 94, NARA.

34. Major General Earl Van Dorn.

35. The raid occurred March 25, 1863. Forrest's cavalry burned the railroad bridge at Brentwood, Tennessee, and captured some 750 officers and men belonging to the 22nd Wisconsin and 19th Michigan. *OR,* vol. XXIII, pt. 1, p. 186.

36. The Union army's most famous and effective mounted infantry organization was the so-called "Lightning Brigade," created early in 1863 by Colonel John T. Wilder, and composed of Indiana and Illinois troops. See Richard A. Baumgartner, *Blue Lightning: Wilder's Mounted Infantry Brigade in the Battle of Chickamauga* (Huntington, W.Va.: Blue Acorn Press, 1997).

37. Colonel William H. Bennison, 78th Illinois, resigned September 2, 1863. *Illinois AGR,* vol. V, p. 3.

38. Brigadier General Absalom Baird.

39. Private Jonathan Dilley died March 28, Private Joseph Andrews March 30 and Private David Jack April 1. *Ohio Roster,* vol. VIII, p. 426, 427.

40. Van Dorn attacked Franklin from the south April 10, 1863. His repulse was aided partly by the firing from Fort Granger of two 24-pounder siege guns and two 3-inch ordnance rifles belonging to the 18th Ohio Battery. *OR,* vol. XXIII, pt. 1, p. 223.

41. Captain William E. Merrill graduated first in the U.S. Military Academy class of 1859. Francis B. Heitman, *Historical Register and Dictionary of the United States Army* (Washington: Government Printing Office, 1903), vol. I, p. 705.

42. Colonel Louis D. Watkins, 6th Kentucky Cavalry (U.S.), was promoted to brigadier general in September 1865. Warner, p. 544.

43. General Granger wrote of the April 27 expedition: "This daring feat shows what our cavalry is made of. The surprise and capture was made almost immediately under the eyes of Van Dorn, within 1 mile of his main body." *OR,* vol. XXIII, pt. 1, p. 322.

44. Van Dorn was assassinated May 7, 1863 at his Spring Hill headquarters by a local citizen. Boatner, p. 867.

45. At the time, Colonel Daniel McCook Jr.'s brigade of four regiments was stationed at Brentwood.

46. The U.S. Christian Commission.

47. At the time, Brigadier General Green Clay Smith commanded the 4th Cavalry Brigade, Department of the Cumberland.

48. Captain Edward P. Bates, Company C, was a native of Hartford, Trumbull County. Edward P. Bates CSR, RG 94, NARA.

49. On May 18, Major General Ulysses S. Grant's Federal army began investing the Confederate stronghold of Vicksburg, Mississippi, whose defenders under Lieutenant General John C. Pemberton found themselves bottled up and besieged. Union assaults May 19 and 22 against Vicksburg's fortifications, however, were bloodily repulsed. E.B. Long, *The Civil War Day by Day: An Almanac 1861-1865* (Garden City, N.Y.: Doubleday & Company, 1971), p. 354-356.

TWO

'*If you love your country,*
aim low — aim well'

On June 8, 1863 the 125th Ohio was transferred from General Gordon Granger's command to the 3rd Brigade, 1st Division, 21st Corps. The brigade, led by Colonel Charles G. Harker, consisted of the 64th and 65th Ohio, 3rd Kentucky and Opdycke's eight-company regiment.

Two weeks later the 125th joined Harker at Murfreesboro, from where Rosecrans' Army of the Cumberland began a series of movements that maneuvered Confederate General Braxton Bragg's forces out of middle Tennessee, all the way to Chattanooga. Rosecrans resumed his advance in mid-August, crossing mountainous terrain in three columns with hope of severing enemy supply and communication lines below Chattanooga. Although Bragg abandoned the strategically located town early in September, he found Rosecrans' widely separated columns inviting targets to attack in detail. With the imminent arrival of reinforcements, especially General James Longstreet's corps transported by rail from the Army of Northern Virginia, Bragg launched his main assault September 19 nine miles south of Chattanooga near the banks of West Chickamauga Creek. The ensuing battle of Chickamauga was among the war's bloodiest. In its first major fight the 125th Ohio, Opdycke boasted, "behaved as gallantly as men could," and earned for itself a celebrated nickname.

Camp near Triune, Tenn.,
June 3.

Dear Chronicle: After a sojourn in and near Franklin for nearly four months, or since the 12th of February, the order has been given, "Forward." Accordingly at 5 o'clock yesterday morning our Division commanded by General Gilbert left the old camp, near

Franklin, and marched to this place, a distance of twelve miles. We arrived early in the afternoon, pitched our tents in a beautiful field covered with nice green grass, and were soon enjoying the rest which tired soldiers know so well how to appreciate.

There had been a good shower the day previous, and the dust which is generally so annoying to troops was completely settled. Our route lay through the most beautiful country that I ever saw. Large fields of wheat and corn were to be seen at all times during the march. The wheat will soon be ready for harvesting. It looked very promising. The corn was late; just large enough to commence hoeing. There were a few negroes at work in the fields, but the most of them have "done gone and run away," leaving their masters to get along as best they can. I was surprised to see that there was so much land under cultivation, but I think that there is no danger that the rebel soldiers will be benefitted by the grain that is raised in this section.

The fruit crop will be large; peaches are nearly full grown, and raspberries beginning to get ripe. We have been thinking of petitioning Uncle Sam to divide up the large farms of these secesh owners and give them to us to settle on when the war is over. What right — either legally or morally — have these traitors to one, two, or even three thousand acres of the finest land in the Union? And then if we soldiers are allowed by Uncle Sam (we shall not ask any one else) to settle in this country, there need be no fears of another rebellion. We will teach these people the true and fundamental principles of our government, and in a short period of time show them by our school houses and churches, our workshops and our superior skill in the arts, the great difference which exists between the land of the free, and the land of the slave.

The Union soldier feels that he is fighting for a great principle, and that if he fails to establish that principle, then has all his labor been in vain — then is his having been a soldier been in vain, then has all that has made us a great and enlightened people been lost to us and to the generations which follow after us, and feeling thus, he toils on earnestly and patiently, knowing that his reward lies at the end.

We are expecting to march again tomorrow, and I presume will soon be more closely connected with the grand Army of the Cumberland. Look for stirring news soon; for when Rosecrans moves it means something. Colonel Opdycke grows more popular with men and officers every day. We have the utmost confidence in his skill and daring as a leader. Our Regiment is now in good fighting condition. The sick have all been sent to Nashville.

Yours,   W.

— *Western Reserve Chronicle,* June 17, 1863

Camp Hillsboro, Tenn.,
July 10.

Dear Chronicle: So incessant has been our perambulations of late that I have found neither time nor opportunity to write to you.

After Gilbert was relieved of his command at Triune (to the infinite satisfaction of all concerned), we were hurried to Murfreesboro. Rosecrans was ready to move to the front, and two days for rest and preparation was all that was allowed us to fit ourselves for the march. We were assigned to Wood's[1] Division, 21st Army Corps, Major General Crittenden[2] commanding. Heavy rains set in the day we started [June 24], and have continued with very little intermission since. Our Corps, comprising the extreme left of the army, took the Manchester road. We made good time until we arrived at Bradyville, where the stone pike terminated and we were forced to take the mud. Our long wagon train soon became involved to an extent which defied the combined strength of our disheartened mules. A regiment at first, and finally a whole brigade, was detailed to assist, but all would not do; our tardy movements were like to defeat the purpose which we had started out to accomplish, and a remedy must be applied. Orders came to burn the baggage, tents, messboxes, blankets, trunks and cooking utensils, and they were hurled in one promiscuous mass into the flames. Even hand-trunks were destroyed, leaving many officers without a change of clothes. Ammunition and provisions were alone saved.

Again we pushed forward with renewed vigor, splashing through the mud and creeks, clambering over logs and crouching through brush and briers. The rebels everywhere fled at our approach. We pursued them through Hillsboro, Pelham and as far as the mountains.

Since the 26th of June, two days after we left Murfreesboro, we have been on half rations. The roads have been extremely bad, and it has been extremely bad that it has been impossible to get supplies hauled. As a consequence, the country through which we have passed has suffered much. Wheat, corn, bacon, potatoes, chickens, and everything that men or horses could subsist upon have been appropriated. We are now pleasantly encamped at this place, awaiting the arrival of our supplies. Deserters from the enemy come into our lines daily. They represent that Tennessee troops will not follow Bragg out of the State, and say that the mountains are full of deserters. We have very little idea where Bragg has taken up his headquarters, but hope to have an opportunity to institute a search soon.

The country through which we have passed bears many evidences of the evils of war. The woe-begone expression depicted in the countenances of the inhabitants, however, indicates more certainly how tired they are of the present state of affairs, and how much they long for the pleasures of other and better days.

The last salt brought to this place previous to our entrance was procured thro' a committee of old men sent for that purpose. They succeeded in getting three pounds for each family at an enormous expense. Four dollars a pound was paid for coffee and three dollars per pound for sugar. Truly this rebellion has been a luxury to the South!

Colonel Opdycke still continues in the highest favor with the men. His fine military bearing is remarked wherever he goes, by both officers and men. During our hard march through the mud and rain he might at times be seen carrying before him not less than five guns — at other times he would be walking, while a fatigued officer or soldier would be riding his horse. We would that there were more men in the army like Colonel Opdycke.

The health of the regiment was never better than now. Late good news from Vicksburg and the East has greatly elated us.[3] The prospect of a speedy and successful termination of the war is the constant theme of discussion. Through the suffering and toil of a few more months, we see the pleasures of home standing forth in fond relief. Not that we are homesick, for we are not; but we desire peace more than war. Yet, tired as we are of this most cruel and unnatural rebellion, we would not surrender a principle or yield a point in any way tending to diminish the former glory of the Republic.

Sergeant Major Smith[4] and Orderly Sergeant N. Phillips,[5] of companies B and C, started for home day before yesterday for the purpose of recruiting for the regiment. They are both highly deserving, faithful soldiers and we wish them good success. Recruits

Trumbull Countian Nyrum
Phillips rose from Company C
sergeant to regimental
quartermaster and adjutant,
finishing his service as an aide
on Opdycke's brigade staff.
While in the army he lost his
mother and two brothers
to fatal illnesses.

Clark, *Opdycke Tigers*

from old Trumbull will find a hearty welcome in the 125th. Come
on boys, and let us close this war up strong.

Respectfully yours,
CEYLON

— *Western Reserve Chronicle,* July 22, 1863

Hillsboro, Tenn., July 27, 1863.

Eds. Herald — Not having seen any correspondence in your
paper for some time past from the 125th Regiment, I make free to
post you on our whereabouts, &c., and of our participation in the
late successful advance of the Army of the Cumberland.

On June 21st, in company with the 124th Ohio, we left General
Gordon Granger's command and marched to Murfreesboro, a distance of twenty miles. On our arrival at that impregnable locality,
the 124th and 125th separated, having been more or less in company ever since their departure from Camp Cleveland; they having been assigned to Palmer's[6] and we to Wood's Division of the
21st Army Corps, which is the First Division. We were brigaded
with the Third Brigade, commanded by Acting Brigadier General
Harker, Colonel of the 65th Ohio, and now composed of the 3rd

Kentucky, 64th, 65th and 125th Ohio Infantry and the 6th Ohio
Battery. We had hardly rested from our Triune march, or had
time to look at the elephantine fortifications at Murfreesboro, ere
we were, with shelter tents strapped upon our backs, again pedes-
trianating Dixieward.[7]

The first morning (24th June) of the advance was one which
foreboded melancholy meteorologically, but the first few drops of
rain were gladly welcomed by us, as it effectually laid the thick
white dust which coated the ground's surface — rain being at any
time more welcome on a march than oppressive heat and choking
dust. We scored ten miles, and, amidst a most pitiable rain,
pitched our tents for the night. During the afternoon we heard
heavy firing in a southwesterly direction (Beech Grove), and anticipated hot work ahead.

On the following day the 125th were detailed to guard the ammunition train — guarding trains is at all times tougher footing
than marching with the main column, especially on such roads
and in such weather as we had then. We proceeded another ten
miles, and again halted. On Friday, the 26th, we did not move.
On that day the sun showed himself a short time, when, availing
ourselves of the opportunity to "make hay," dried our ammunition.

Saturday, the 27th, was the hardest day's march we experienced; not from the distance traveled, but from the wretched
ground traveled over, and having to climb a very high hill and
help our baggage trains up; all this under the most *incessant*
rains conceivable. On Sunday, 28th, we encamped one mile north
of Manchester, and on Tuesday, 30th, moved here, where Crittenden's and Thomas'[8] Army Corps had concentrated.

Towards dark on the evening of Wednesday, July 1st, General
Wood's Division moved out on the Hillsboro road. Passing beyond
that village in the darkness of the night, we waded the swift current of Elk river and bivouacked. At daylight on the 2d, we were
again on foot, and at noon reached an insignificant place called
Pelham, which is to the right and rear of Tullahoma, but it was
too late for a fight, as the greater part of Bragg's army had escaped across the Tennessee, and our getting in the rear of Tullahoma did not avail much. It being then known at headquarters
that Tullahoma was in our possession, General Wood only gave us
time to rest and cook some dinner, when he returned to Hillsboro,
but on Friday, 3d, we again countermarched, and returned to
Pelham, where we remained until Wednesday, 8th inst., when we

again returned to Hillsboro, where we have been encamped ever since.

From June 29th to July 19th, we were on short rations, most of the time making one day's rations of coffee, sugar and hard tack last three days. We are all in good spirits and anxious for another advance. The news from other armies has been cheering and encouraging.

To-day one of our scouts handed me a Chattanooga Daily *Rebel* of July 7th. I enclose the treasonable document for your inspection. I will write again in a few days.

CATO.

— *Cleveland Herald,* August 4, 1863

Hillsboro, Tenn., August 2d, 1863.

Dear Chronicle: All is quiet with the Army of the Cumberland. The rebels ceased to molest us. We perform picket duty, scout and forage about the country without a show of opposition. We see butternuts[9] quite frequently it is true, but they are deserters from Bragg's Army. They inform us that Tennessee soldiers would not leave the state to follow the desperate fortunes of Bragg, and as a consequence, the mountains are full of deserters.

Tennessee is lost to the Confederacy and the people see it if Jeff Davis does not. Those who, one year ago, were the loudest brawlers for secession, are now the most inoffensive people you could imagine. With a subdued air they implore our soldiers not to take their property. *"I never done nothing,"* they weakly interpose, and one would think from their conversation that they were paragons of propriety. These have-been Secessionists go further; they come to our Generals with recommendations from neighbors and friends, equally guilty with themselves, asking protection; and, what is most remarkable, protection papers are in a plurality of cases granted them. To this same class of men Government receipts to the amount of thousands of dollars are given by our Quartermasters for forage. To us who are acquainted with the duplicity of these scoundrels it is perfectly clear that not one receipt should be honored by the Government, but that through the craft of wily lawyers and agents every cent will be collected after the war, there is at least strong probability.

The progress of John Morgan through Indiana and Ohio was watched with the most intense interest by our soldiers in this Department. We began to tremble for the safety of our friends, and some even suggested the propriety of having their families sent on to them when we heard of his capture.[10]

We are still looking forward with some interest to the coming draft. We are ready to welcome with a hearty good will a few score of our Trumbull friends who are fortunate enough to draw the lucky ticket. The ranks of our noble regiment need recruiting and we trust we shall not be forgotten when the fortunate hour arrives. We say to our acquaintances pay no three hundred dollars, nor hire a substitute, but come right along and we will close the war up strong.

Respectfully yours,
CEYLON

— *Western Reserve Chronicle,* August 12, 1863

Hillsboro, Tenn., Aug. 10, 1863.

Eds. Herald — Here we lie, quiet and inactive, as though there was no foe before us, no battles to be won, no rebellion to be crushed. For more than a month we have occupied the same beautiful encampment. We have drilled occasionally, and performed picket duty as though we were every day in danger of an attack. But not a reb has ventured to approach us save as a suppliant for mercy.

Poor Bragg! His forces disheartened and disorganized, can no more be urged to the fierce attack. While at Franklin and Triune, Van Dorn and Forrest's cavalry were nightly prowling about our lines, and we were almost daily drawn up in line to resist their bold dashes. Where are they now? They have long since ceased to molest us. Our forage trains, under comparatively small guard, penetrate far into the country with impunity. If we meet any butternuts whatever they are such as are tired of secession and are willing to accompany us to headquarters where they may take the "oath."

Jeff Davis can no longer rely on Tennessee soldiers. They will desert his false standard at every good opportunity. The mountains are full of refugees and deserters from Bragg's army, as I am credibly informed, who are enlisting under the old Stars and Stripes. Union soldiers are aware of this, and are encouraged. The army was never so confident of speedy success as now. A few more *victories* and the contest is over — the day is won.

We look to the draft now taking place in the loyal States to fill our decimated ranks; then, with an unbroken front, we shall again be ready to lead the charge. If we are properly reinforced, next Spring will see us returning to our homes; but if no more help is to be sent us, if the heroes of two winters are to be the veterans of a third, if we are to fight while our friends ignobly skulk behind the delusion that their duty to the country is performed when three hundred dollars are paid and no life jeopardized, then may the war continue indefinitely.

Notwithstanding the enemy is beaten back on all sides and success everywhere attends our arms, our numbers are every day becoming less; we are continually pushing further from our supplies, and large forces are needed to protect our lines of communication. In view of these facts we appeal for help. We say to brothers and friends, *come.* Let every man that is able, shoulder a gun and prove his patriotism by using the strength God has given him in the field to crush the common foe.

The election in Kentucky has met our fondest hopes. We shall expect a similar but more decided victory in Ohio. Vallandigham[11] has not a friend in the army. Though we are in favor of free suffrage and a free expression of opinion, I do not think it would be safe for a man to espouse the cause of the traitor candidate for Governor of Ohio, in this Department. We are willing to *fight* traitors, but to *vote* for them or *endure* them in our midst, would be too great a tax upon our generosity.

The 125th was never in better fighting condition than now. I have heard it remarked repeatedly by outsiders that our regiment is the best drilled and best disciplined of any in this Division. This I say to gratify no feeling of vanity, but in justice to Col. Opdycke, to whose personal effort is due the good name we bear. His fine military bearing and manly conduct win for him the esteem of officers and men wherever we go.

From present indications the Army of the Cumberland will again be on the move within ten days. The time we have spent in camp apparently inactive has been used to a good purpose. The men have been newly clothed, supplies have been accumulated in large quantities, so that everything is ready for a permanent advance when the order shall be given.

Respectfully yours,
CEYLON

— *Cleveland Herald,* August 17, 1863

Hillsboro, Tenn., Aug. 15th, 1863.

Dear Chronicle: Yesterday, in company with four others, with an ambulance and two good saddle horses, I went twelve miles from camp to the foot of the Cumberland Mountains, in quest of vegetables. We took nothing but our revolvers with which to protect ourselves, and to quiet any anxiety that might arise in the mind of any who may read this letter, I will here state that we met with no misfortune. Not a reb obstructed our field of vision.

As articles of barter we took with us a few pounds each of coffee, sugar, salt and soap. Greenbacks are in good demand through the country, but as the above articles have been very scarce for a long time, they are especially acceptable. We had no difficulty in exchanging two pounds of coffee for one bushel of potatoes, a pound of sugar for four quarts of tomatoes, one pound of coffee for two pounds of butter or two dozen eggs, pound of salt for a nice young chicken, and other things in proportion. Green corn was plenty, so we helped ourselves without price. Although this is an excellent fruit country, apples and peaches are neither plenty nor of good quality this season. We secured about two bushels of each at *moderate* rates. Having our load complete, we returned to camp in time to pass the pickets before the countersign was put on.

As on other occasions, I improved every good opportunity of observing the tendency of public opinion concerning the rebellion. One thing is perfectly established, the mountains are full of deserters from the rebel army. Deserters come to our lines daily, anxious to swear allegiance to the Union. Many intelligent men declare their conviction that a fair election would place Tennessee back in the Union by an overwhelming majority. Frequent mass meetings held by the citizens wherever they can have the protection of our army, and where none but Union sentiments are expressed, is another evidence which cannot be mistaken. If Rosecrans' army is not doing much with sword and bayonet to crush the rebellion, it is yet doing good work. Wherever we go a love for the Old Union immediately begins to develop. We leave a loyal people behind us.

Our sick are being sent to the rear as fast as possible, our trains are being loaded with ammunition and supplies, and every indication points to an onward move very soon. When our ranks shall be swelled by the addition of thirty thousand more loyal men from Ohio, what a sweep we will make through the cotton-fields of the Gulf States. Again we say "come on boys, and see the country!" Don't rust out at home while so wide and so noble a sphere of action is opened to you ....

Respectfully yours,
CEYLON

— *Western Reserve Chronicle,* August 26, 1863

Thurman, East Tennessee,
Aug. 22d, 1863.

Eds. Herald — Sunday morning last, 16th inst., while brushing up for the usual inspection, welcome marching orders came. Our little shelter tents were struck, when we left the prettiest and neatest camp we have yet had. Soon after we left Hillsboro, dark clouds portended another rainy march, but it was only a passing storm which laid the dust and cooled the atmosphere for the rest of that day.

Sunday night we camped near Pelham and on the following morning moved to the foot of the mountain, where we unloaded the teams of half their baggage and ammunition. Clambering up the mountain with our own personal baggage and equipments, we stacked arms above, rolled up our sleeves, and headed by our Colonel we descended about midway; and with stout ropes prepared to help up the teams. Every man seemed cheerful and anxious to put his shoulder to the wheel. Our officers were especially energetic, cheering up the men and taking active hold themselves. Generals, Colonels, Captains and high privates were promiscuously mixed tugging away. Col. Opdycke worked hard as he always does. Taking his position at a precipitous point, he rendered great and valuable aid, directing contrary mules, repairing the road, &c., &c.

The night was fine but dark. Blazing fires at intervals of twenty paces illuminated the ragged, rocky path the whole distance. Teamsters' oaths mingled with troops yelling and cheering; officers shouting, commanding, directing, &c. Bands playing patriotic music filled the programme with nocturnal harmony. Long after midnight the scene lasted. As soon as the last wagon reached the top, the descent of the unloaded wagons commenced. Meantime the men rested. At dawn we were again at work dragging up the same train with the remaining baggage. At noon, with triumphant yells, "a good pull, a strong pull, and a pull altogether," we again dragged up that last wagon. Notwithstanding our heavy night and morning's work, we immediately marched to Tracy City, where a branch from the Nashville and Chattanooga railroad terminates, connecting with important coal mines near that settlement. Before daylight on Wednesday morning the 1st[12] and 3d Brigade took up the line of march from Tracy City, and after dark commenced the descent, accomplishing a distance of twenty-eight miles that day. We now rest in the Sequatchee Valley, and on the river of that name. Herewith I send you General Wood's congratulatory order.

CATO.

General Orders No. 71   Headquarters, 1st Division,
Thurman, East Tennessee,   August 20th, 1863.

The Commanding General congratulates the troops on their splendid and successful march from Hillsboro and Pelham to this place. The order to march was received at daylight, Sunday, the 16th inst. In it the distinguished and gallant Commander of this army appointed Wednesday evening, the 19th, for the Division to be at this place. To do so, with heavily laden transportation, had to march 50 miles across one of the principal mountain ranges of our great country. The line of march required the troops to make an ascent and descent of twelve hundred feet. The ascent required forty-eight successive hours — the severest labor — the men by day and by night dragging the transportation up the steep mountain road by hand. This toilsome labor they performed, not only willingly but cheerfully. In overcoming the difficulty of the ascent, more than one-half the time given the Division to reach this place was consumed. This result imposed the necessity upon the two Brigades which marched from Hillsboro of accomplishing twenty-eight miles on the fourth. They did it in good order.

Such energy in overcoming obstacles, such capacity of endurance of the severest toil and fatigue, and such cheerfulness and enthusiasm in all these labors and privations, render the troops who exhibit them invincible, and will always crown their efforts with success.

The commanding General hails the result of the march just concluded as an auspicious omen. The campaign *must* be successful. Chattanooga will be ours.

By command of Brig.-Gen. Wood.
(Signed)   M.P. Bestow,
Captain and A.A.G.

[Official.]
Alex. Moffatt, Lieut. and A.D.C.

— *Cleveland Herald,* August 29, 1863

Brigadier General Thomas J. Wood, a native Kentuckian, Mexican War veteran and U.S. Regular officer, enjoyed "universal confidence" among his division subordinates prior to Chickamauga. Opdycke thought Wood was a "delightful conversationist."

Sequatchee Valley, Tenn.,
August 26th, 1863.

Eds. Chronicle: Our Division under command of General Wood left our beautiful camp at Hillsboro on Sunday, the 16th inst. It was the most convenient and by far the pleasantest camp that we had ever occupied, and we felt like bidding a dear friend good bye when the bugle sounded "strike tents" and we marched away.

We could not see the propriety of commencing our march on Sunday, but for some reason or without any reason, that day seems to be chosen in preference to any other for the commencement of any great march. We encamped the first night at the foot of the mountain, and the next day was occupied in climbing to the top, a distance of perhaps two miles. At places the road was so steep that the men were obliged to assist the mules in dragging up the heavily loaded wagons. This was accomplished by fastening long ropes in front, and the men taking hold on either side, when the united strength of men and mules would soon take them up. It was severe work, and was not entirely performed until the morning of the next day. It reminded us of "Bonaparte crossing the Alps," though of course we felt that it was comparatively a small affair. Still to us who have never traveled "it was considerable," and we shall mark it down as quite an episode in our journeyings through Dixie.

We marched some thirty-five miles on the mountain before we began to descend. The roads were generally good, though many miry places were passed through only by help of the men lifting at the wheels, or steadying the wagon to prevent it upsetting in rough stony places. It was not an uncommon sight to pass broken wagons; wheels sometimes minus spokes. But they were soon repaired, reloaded, and again on the march. Enough ground was under cultivation to furnish us with roasting ears and potatoes. Some of the boys declared that they would not sleep on the ground, but weariness and want of sleep would overcome the fear of snakes.

The third day our ears were greeted by the whistle of a locomotive. It took us completely by surprise. We had seen the road before but did not suppose that it was in running order. Thus you see that not even the mountains prevent the Iron Horse from following the march of our victorious armies. The fourth day we marched twenty-eight miles and all reached the foot of the mountain with the exception of a part of our wagon train. Our Regiment passed over the last five miles in one hour. The boys will not soon forget their "double quick" down the mountain with Col. Opdycke on foot and at the head leading them on.

Went into camp at eight p.m. having marched over fifty miles including the ascent and descent of a mountain 1200 feet in height, all of which was accomplished with great cheerfulness by all. We are now twenty-six miles from Chattanooga in a very nice valley, and are feasting on peaches, apples, potatoes, honey, &c., which abound here in endless quantities. Inhabitants all secesh, consequently we confiscate anything we want. We are all in fine spirits and confident of success. Gen. Wagoner[13] [sic] is reported within fives miles of Chattanooga. We are expecting to move at any moment.

Yours,　W.

— *Western Reserve Chronicle,* September 9, 1863

Chattanooga, Sept. 25th, 1863.

Eds. Chronicle: Anxious hearts all over the land are beating for loved ones far away who have participated in one of the most fearful battles ever waged on this continent; with eager, yet with trembling hearts, their eyes will glance over the list of casualties — the killed, the wounded and the missing. May God help those who read the name of a father, a son, a husband, or a brother.

It is with mingled feeling of pride and sadness that I attempt to describe the part which our regiment has taken in the battle. Of pride, that I am a member of the regiment which has gained for itself a name second to none among all those who are now so nobly defending the institutions of our country; of sadness, that the names of so many brave soldiers will be placed on the list of casualties.

The battle commenced on the afternoon of the 18th, the battle line being about ten miles beyond Chattanooga, the centre resting on Chickamauga Creek. On the 19th, the roar of artillery and musketry was almost deafening, the rebels having attempted to turn our left wing, and get between us and Chattanooga, but it was on Sunday, the 20th, that the battle raged the fiercest, and our regiment suffered the most. A portion of Lee's army under command of Longstreet,[14] in column eight deep, made repeated charges on our troops, and were each time gallantly repulsed.

Prisoners told us, afterwards, that we did not fight like the Eastern troops, that they could always drive them back when they
made such charges. But our boys know not the word retreat, or
run, except to the right or left along the line, and when the lead-
en hail rained heaviest, they still fought and bravely kept their
faces toward the foe, while many a brave heart fell to rise no
more. It seems that the whole rebel army is in front of us, and
that they are staking their all upon the result of this contest. We
have fallen back towards this town, are strongly entrenched,
awaiting reinforcements. We were enough for Bragg, but when
Lee steps in, then we call for more help ....

General Garfield[15] witnessed our fighting for about two hours,
and Gen. Rosecrans has since personally paid us the following
tribute of praise: "I thank you sincerely for the magnificent manner in which you fought."[16]

But the battle is not yet decided. We still hear the roar of artillery, and the "Assembly" for the infantry to "fall in" may sound
at any moment.

W.

— *Western Reserve Chronicle,* October 14, 1863

Camp near Chattanooga, Tenn.,
Sept. 26, 1863.

Eds. Herald — Under the press of duties which have devolved
upon me since the great and terrible struggle through which we
have passed, I have been unable, until the present evening, to
obtain sufficient leave to apprise our friends of the part which the
125th took in that terrible and bloody action. Nor do I now attempt a description of the awful grandeur of that bloody field
where a few of America's freemen met the combined horde of rebel troops, but shall simply narrate such facts as came under my
own observation and hearing.

For some days previous to the 19th inst., the first day upon
which the great battle of Chickamauga was fought, our division
held Gordon's Mills and the line of the West Chickamauga River,
being the extreme left of the Army of the Cumberland. On the
18th the enemy made a reconnaissance along the line of the
whole army, which was promptly resisted. But the immense
clouds of dust that arose upon the left clearly indicated that it
was that point which they had determined, if possible, to destroy.
During Friday night Thomas' Corps and part of Crittenden's

Gordon's Mills Ga.

"Severe battle on our left. 3d brig. ordered out
at 1 P.M. Got in rear of reb line and chased
a number of regts. I took 3 prisoners, Adjt 3
& others 3. 125th lost 1 killed & 11 wounded.
Firing severe. We were surrounded. Harker
put me in comd of 2 regts & complimented me
for getting them out safely."

• Emerson Opdycke diary, September 19, 1863

---

moved beyond us to the left, and McCook's [17] forces started for the
scene of action. Saturday, the 19th, came in foggy, preventing an
early onset.

Thousands of armed men lay within a few hundred yards of
each other, still as the silent tomb at midnight hour, awaiting the
morning sunbeams to clear away the fog and light them to the
roar and clash and death of bloody battle. At about 10 A.M. the
awful silence was broken by the crash of musketry and the deafening roar of artillery. Soon the battle raged most terribly, nor
ceased till night had drawn her sable curtain over earth, shutting
out the sight of foe from foe. At 1 P.M. our division was ordered to
support the left. We entered the field on the double quick and
were soon baptized in blood. This was the first general action that
the 125th had ever been in, but cheered on by the cool and noble
daring of her Colonel, she could but conquer.[18]

Just as we entered the battle, Col. Opdycke turned to us and
said, "Men of the 125th Ohio, if I or others fall, *stand in the ranks
till victory is ours.*" And while the battle was raging at its greatest fury, he forgot not to cheer his men, for loud above the din of
conflict arose his voice, "Now men of the 125th, if you love your
country, aim low — aim well." Side by side and shoulder to shoulder did the men of the 125th contest that bloody field, aiming low,
and promptly obeying orders, for our Colonel had taught us that
in these consisted our safety. We took nine prisoners,[19] three of
which the Colonel captured himself, while our brave Adjutant,
E.G. Whitesides, captured three of the others.[20] The rebels were
routed in confusion, and after dark we bivouacked on the field of

battle, without fires, without supper, many of our men without blankets, while the cold north wind chilled our wearied limbs almost to numbness. At 2 A.M. of the 20th we were again called up to arms and having marched a couple of miles to a new position, we halted, a hasty breakfast was prepared and eaten, and we then took our place as a reserve near the left of McCook's and at the right of Thomas' corps.

The battle opened on Thomas' left at about 8 A.M., and so severely that our division was ordered to move on the double quick to support it. McCook was ordered to close, by a left flank movement, the gap between his left and Thomas' right, which had been made by *our* moving to the left. McCook failed to perform his part of the great plan, which the enemy were not long in discovering, and forced a heavy column of Longstreet's corps through the gap, attacking us on the right flank and rear almost at the same instant we were encountering the enemy in front. This destructive cross fire almost annihilated the 1st Brigade of our Division, and the 2d Brigade being in Chattanooga left ours (the 3d) to contend almost alone with an entire corps of the enemy. This movement of the enemy cut McCook's force almost entirely out of the fight, and but few of his men returned to solid cooperation. It is said that McCook himself was soon after seen in Chattanooga, a distance of ten miles from the battle.

So soon as we became aware the enemy had got behind us, we changed front to rear on our left, and found ourselves face to

Opposite: Edward G. Whitesides, the regiment's first adjutant, was a pre-war express agent living in Pittsburgh. His diary for September 19, 1863 reads in part: "At 1 P.M. we were called upon to reinforce the left. We advanced and engaged the enemy in a dense woods at about 2 P.M. Captured one rebel officer and eight enlisted men, with a loss of the 1st Sgt. of Co. A [James B. Morris, killed] and eleven men wounded. The regiment was under fire until after dark. Lay on our arms without fires until midnight when we retired a mile and a half towards Chattanooga and camped on the side of a high hill, and lay down before large fires for a few hours rest."

As a captain in 1864-65, Whitesides served in several brigade and division staff positions, was severely wounded in the left thigh at Kennesaw Mountain, and mustered out with Company A. He received a major's brevet in 1867. Carte de visite by N.E. Lewis Photographer, Cleveland, Ohio.

face with a line of the enemy that stretched far beyond us on both right and left. The enemy knew that if they could crush us and defeat Thomas' left, they would thus hold the ground between our army and Chattanooga, thereby insuring the destruction of the Army of the Cumberland, and made the most desperate efforts to push us from our position. Gen. Wood and Col. Harker braced up the lines for a most deadly and unyielding struggle. On came the enemy in solid lines of eight men deep, and we soon stood face to face with the boasted veterans of the Richmond army. The musketry firing became most terrific, and raged with an unparalleled violence. At this moment Gen. Thomas ... visited us in person, and said to Col. Opdycke, "Colonel, this point *must be held.*" Col. Opdycke's reply was, "We will hold this ground or sleep to-night in Heaven."

Gen. Wood ordered the brigade to advance upon the enemy. Col. Opdycke having commanded fix bayonets, rode to the front of his regiment, and turning to us said, "Men, I will lead you, follow me," and plunged into the midst of the foe, followed by the entire brigade. The regiments upon the right and left of us, however, could not keep pace with the lightning speed at which we advanced, and were left in our rear. The whole rebel line fled in confusion, being unaccustomed to the resistless charges of our western troops, and terrified by the glitter of our cold bright steel. We halted and lay down behind a fence while the other regiments came up, and prolonged the line to our right and left, the right resting on an eminence some fifty feet above us.[21] Gen. Wood came up to Colonel Opdycke and said, "Colonel, that charge of the 125th was a most splendid thing."[22]

The enemy now advanced their second line. We could not but admire the terrible splendor of their advance, the men closing up their ranks with all the coolness and precision of a drill, and stretching far beyond our right and left, seemed confident that they could crush us at a blow. Their fire soon swept over us, while our boys hurled back the leaden storm, and swept their ranks in terrible destruction. Just at this moment the regiments on our right and left retired,[23] and but for the cool command of our Colonel, and steady obedience of the men we would have been annihilated.

The regiment on our right retiring, led our right company [A] to believe that there had been a general order to retire. They arose and faced about, when the Colonel ordered them to their

Erroneously reported killed, Second Lieutenant Martin Van Buren King, Company G, temporarily commanded Company C at Chickamauga before he was seriously wounded and captured September 20. Disability forced his resignation five months later.

Clark, *Opdycke Tigers*

post. Accustomed to obedience, they resumed their place and side by side with their companions fought the foe. We held the ground alone until the rebel lines were on a prolongation with our right upon the hill, when their fire directly enfiladed us, killing Lieut. King,[24] seriously wounding Capt. Yeomans[25] and Lieut. Barnes,[26] and laying many of our brave boys in the dust. It was apparent that we must retire to our brigade or perish, and although we were outnumbered, yet we were not conquered, for as we fell slowly back Lieut. Clark remarked, "They may outflank us and kill us, but whip us they can never." As we were falling back to a better position a flying regiment broke obliquely through our ranks, but even then discipline prevailed, and our boys closed up the ranks as if on morning drill. Col. Harker came up to Colonel Opdycke and complimented him for his bravery and the splendid fighting of his men.

The enemy threw themselves with reckless courage on our lines, and hurled a perfect storm of balls upon our thinning ranks. Front to front and man to man we faced the rebel hordes, resisting all assaults and crushing back the vaunted lines of Longstreet's force. For two long hours a sea of fire swept that awful field, piling the ground with dead and wounded of both friend and foe. To

L. M. Strayer Collection

Sergeant Jacob Jewell, Company F, received a slight facial wound at Chickamauga when his "cheek [was] laid open by the kiss of a minie." A Licking County native and teacher, he was photographed in 1865 at Leeson's Gallery, New Orleans.

destroy *us* was to ruin the entire army; while Gen. Thomas had said, "This position must be held." We could but die, we must not yield, and for long hours we beat back the enemy, almost single handed, and *did hold our own.*

While the battle was raging at its fiercest heat and it seemed no living being could withstand the tide of death, our Colonel rode along the line, and raising his voice above the din of crashing arms cried out, "Stand firm, my boys, I am willing to fight for my country, to die for her, and I hope you are with me." During the entire strife Col. Opdycke remained upon his horse, being the only officer on the field who did not dismount, a conspicuous mark to the enemy's sharpshooters, who soon discovered the bravest on the field. Nobly was he seconded in his efforts by our gallant Adjutant, Lieut. E.G. Whitesides, of Pittsburgh, Pa. It was miraculous that they escaped. The Colonel had a ball shot through his blouse, slightly wounding his shoulder. His horse was wounded several times by the rebel marksmen. The Adjutant had his horse shot from under him, but mounted a second. Said an officer to Colonel O, "Sir, you must have a charmed life, for I cannot see how you could live in such a storm of lead." Generals Thomas and Garfield, in speaking of this part of the battle, said it was the grandest repulse they had ever seen.

At three P.M. the 41st Ohio came up and lay down a couple of rods in rear of our regiment just as the enemy made his last and most terrible assault.[27] A mutual recognition immediately took place between our Colonel and the 41st, for he had formerly served as Captain in their regiment. Three thundering cheers arose for Col. Opdycke from their ranks, while he, hat in hand, amidst the cloud of balls, sat upon his horse and commanded his regiment. Two pieces of artillery were placed under our Colonel's command, with which he swept the rebel ranks with grape and cannister.[28]

We remained upon this spot until all firing had ceased and night had closed upon the bloody field. Our Division, after being ordered the second time, fell slowly back and without interruption, to Rossville, where General Thomas put the army again in position. It is needless for me to comment upon the heroic actions of the 125th during two days of the most terrific fighting ever recorded, for abler judges than I have spoken of it. General Rosecrans visited our lines the other day, and as his staff arrived near us, Gen. Garfield pointed to Colonel Opdycke and said, "General

"Terrific battle all day. on Barney almost
every moment. No other field officer on
horse in the Division. Changed front
several times under fire. Lead a charge
& routed the enemy then retreated to a
better position, held it & repulsed
3 desperate charges of the enemy.
Thomas & Garfield saw us ...."

• Emerson Opdycke diary, September 20, 1863

Rosecrans, there is the man that sat on his horse through all of
Sunday's fight, and these are the men," pointing to the 125th,
"who stood at their post. For two hours I stood and watched them
hold the enemy in check." Gen. Rosecrans then rode up to our
Colonel, who stood near our battle-soiled and bullet-ridden flags,
and said, "I thank the officers and men of the 125th Ohio Regiment for the magnificent manner in which they fought. Do as
well on your present lines and the 'gray back' of this rebellion will
be broken."

This was a proud moment for the 125th. Cheer after cheer
went up for "Rosy" and our brave Colonel. The 125th have met
the enemy and stood the test most nobly. How could they do otherwise when led on by so brave and efficient a commander as Col.
Opdycke. He has proved his words, "That if we would stand by
him he would stand by us, and lead us to glorious victory or an
honorable grave."

Another most noted item to our credit is the almost unparalleled small number of missing. After two days of the most severe
test for soldiers, but *five* were missing.[29] General Wood says of it,
"It is perfectly grand; I do not believe there is another such record
in this entire army, even among old troops." We have done gloriously. We have won honor, and give us Colonel Opdycke for our
leader we will yet cause the rebels to tremble at the name of the
125th Ohio. But our resistance caused us the loss of many brave
men. Numbers fell to rise no more, while many others are among
the wounded.[30] I will give the list of the killed and wounded in our
regiment as I have it corrected by the Adjutant [here omitted].

An unspecified Chickamauga wound did not prevent Private James A. Needs, Company E, from remaining on duty. He was appointed corporal six months before this portrait was made at the New Orleans photographic studio of John Weiss. Note small brass company letter affixed to his frock coat. Seventeen days before he mustered out in September 1865, Needs was promoted to sergeant.

Company F private John Goetz, a Mansfield mechanic born in Germany, suffered a finger injury at Chickamauga, and a second minor wound at Resaca before appointment to corporal June 7, 1864. Carte de visite by Lilienthal, New Orleans.

I have thus attempted to give you some little idea of the part we took in the above contest. We are now behind our lines near the city of Chattanooga, but shall not long await. Look to hear from the 125th Ohio soon again.

Yours in haste,
VICTOR, 125th Ohio.

— *Cleveland Herald,* October 7, 1863

Notes to Chapter Two

1. Brigadier General Thomas J. Wood.

2. Major General Thomas L. Crittenden.

3. Vicksburg, Mississippi was surrendered to General U.S. Grant's forces July 4, and Major General George G. Meade's Army of the Potomac defeated the Army of Northern Virginia at Gettysburg, Pennsylvania, July 1-3.

4. Sergeant Major Seabury A. Smith was a native of Kinsman, Trumbull County. Before enlisting September 26, 1862 in Company B, he was a dry goods store clerk in Warren. Seabury A. Smith CSR, RG 94, NARA.

5. First Sergeant Nyrum Phillips, Company C, was a farmer born in Greene Township, Trumbull County, and prior to his three-year enlistment resided in the village of Mecca. Nyrum Phillips CSR, RG 94, NARA.

6. Major General John M. Palmer commanded the 2nd Division, 21st Corps.

7. Shelter tents first were issued to the 125th Ohio in mid-March 1863. Regimental historian Charles T. Clark recalled: "Each man was supplied with a piece of canvas about five feet square, having a row of button holes and also a row of buttons around the border. Two of them buttoned together and carried over a ridge-pole elevated on prongs about three feet high, with the ends fastened on the ground, made what the boys invariably called a "pup tent," probably because no animal could enter otherwise than upon all fours. Four pieces buttoned together, with an extra one or a rubber blanket across one end, made a home for four men, which, if less commodious than a wall tent, had the advantage of being always at hand, and required no space in wagons, each man carrying his part of the family mansion with his blanket." Clark, p. 48.

8. Major General George H. Thomas commanded the 14th Corps.

9. A term commonly applied to Confederate soldiers, derived from the butternut-colored uniforms many of them wore.

10. Brigadier General John Hunt Morgan with some 2,500 Confederate cavalrymen crossed the Ohio River into Indiana July 8, and five days later entered Ohio northwest of Cincinnati. Traversing southern Ohio by July 19, he intended to recross the river at Buffington Island, but was badly beaten there by Federal forces. Pursued north, Morgan and 364 survivors finally surrendered July 26 in Columbiana County, Ohio. Boatner, p. 568-569.

11. Peace Democrat and noted Ohio "Copperhead" politician Clement L. Vallandigham had been exiled to the South by order of President Lincoln in late May 1863. Traveling by sea to Canada, Vallandigham ran for Ohio governor in absentia against Union party candidate John Brough, and was soundly defeated in October 1863. His nephew, George B. Vallandigham, served in the 125th Ohio as a Company E private.

12. The 1st Brigade of Wood's division was commanded by Colonel George P. Buell, and composed of the 100th Illinois, 58th Indiana, 13th Michigan and 26th Ohio infantry regiments. *OR,* vol. XXX, pt. 1, p. 44.

13. Brigadier General George D. Wagner commanded the 2nd Brigade

of Wood's division.

14. Longstreet personally arrived at Bragg's headquarters about 11 p.m. September 19, and was assigned command of the Army of Tennessee's left wing. Only five brigades of Longstreet's corps reached the battlefield in time to participate at Chickamauga. *OR*, vol. XXX, pt. 2, p. 287-288.

15. Brigadier General James A. Garfield of Ohio was Rosecrans' chief of staff.

16. Rosecrans' congratulatory remarks were made to Opdycke September 23 while his regiment was drawn up in line at Chattanooga. *To Battle for God and the Right*, p. 97-98.

17. Major General Alexander McD. McCook commanded the 20th Corps.

18. The 125th Ohio primarily fought September 19 in the woods east of Brock Field and the LaFayette Road. It entered the battle with 16 officers and 298 enlisted men. *OR*, vol. XXX, pt. 1, p. 707.

19. The captured Confederates likely belonged to Brigadier General Bushrod R. Johnson's provisional division.

20. Edward G. Whitesides' September 19 diary entry confirms that nine Confederates were taken prisoners, but does not mention his own role in their capture. E.G. Whitesides diary, *Civil War Times Illustrated* Collection, USAMHI.

21. This position was in the northern portion of Dyer's Field. The Confederates chased off by the charge of Harker's brigade belonged to Brigadier General Jerome B. Robertson's Texas Brigade and Colonel William Perry's Alabama brigade, of Major General John B. Hood's division, Longstreet's corps. Peter Cozzens, *This Terrible Sound: The Battle of Chickamauga* (Urbana: University of Illinois Press, 1992), p. 407-410.

22. Adjutant Whitesides recalled General Wood's words as "That was a glorious charge and if I live it shall be made official and go into history." E.G. Whitesides diary.

23. The 64th Ohio was on the 125th's left, and the 3rd Kentucky to its immediate right. *OR*, vol. XXX, pt. 1, p. 708. Their attackers belonged to the South Carolina brigade of Brigadier General Joseph B. Kershaw. Cozzens, p. 416.

24. Second Lieutenant Martin Van Buren King, Company G, temporarily commanded Company C at Chickamauga while Captain E.P. Bates acted as the 125th's major. King mistakenly was reported killed on September 20. He was struck in the back just right of his spine, the minie ball passing through a lung and out at the chest. With other wounded prisoners he was taken to a field hospital at Crawfish Springs, and paroled there September 30. A resident of Petersburg in Mahoning County, King was unable to update Opdycke about his condition until two months following the battle: "I am sorry that I could not write to you sooner. My right arm has been so paralyzed that I could do nothing with it. The doctors say it comes from the wound. I have been very sick since I came home. I have not been out of the village & I have to spend about half my time in bed yet in daytime. I was in hopes I would soon recover & be able to return to my brothers in arms but it seems that I am doomed to a long siege of sickness & pain on account of my wound." King resigned March 1, 1864. M.V.B. King CSR, RG 94, NARA; King to Opdycke, November 19, 1863, Emerson

Opdycke Papers, OHS.

25. Captain Albert Yeomans, Company B, was shot through the flesh of one of his thighs. He resigned May 7, 1864. Clark, p. 130; *Ohio Roster,* vol. VIII, p. 422.

26. Second Lieutenant Albert Barnes, Company E, died October 22 from complications of a broken thigh. Albert Barnes CSR, RG 94, NARA.

27. "Victor's" timing of the 41st Ohio's arrival is in error. The actual time was about 5:30 p.m. Cozzens, p. 480.

28. The guns were 3-inch ordnance rifles belonging to the 18th Ohio Battery. *OR*, vol. XXX, pt. 1, p. 235, 708.

29. Opdycke wrote six days later: "Justice demands that the facts in favor of 4 of the missing be officially noted. Two of them had just joined from hospital; 1 had no shoes, and on crossing a burning turf, on the 19th, his feet became so burned that he and the other two, not being able to keep up, were ordered back by their officer. The fourth one was left to take care of Lieutenant Barnes, which leaves the fifth the only case without excuse in the regiment." *OR,* vol. XXX, pt. 1, p. 709.

30. Exclusive of the missing, Opdycke numbered the 125th Ohio's Chickamauga casualties at 100 killed and wounded. *OR,* vol. XXX, pt. 1, p. 709.

THREE

'Men who could make such a charge could storm Hell and take the Devil by surprise'

Between the middle of October and early December 1863, neither the *Cleveland Herald* nor the *Western Reserve Chronicle* carried anything written by members of the 125th Ohio. Former *Herald* employee Henry Glenville had been seriously wounded at Chickamauga. Commissary Sergeant Hezekiah Steadman, the probable author of "Victor's" letters who was near Opdycke's side through much of the Chickamauga fighting, was detached October 21 for recruiting service in Ohio. The *Chronicle's* correspondent, Lieutenant Ridgley Powers, already was on similar duty in the Buckeye State.

During their absence General Rosecrans' Army of the Cumberland was bottled up in Chattanooga, confronted by Confederate lines that stretched in a huge semi-circle from Raccoon and Lookout mountains to the bank of the Tennessee River north of Missionary Ridge. In the wake of Chickamauga, Rosecrans was relieved of command and replaced by Major General George H. Thomas. His troops in the region were greatly reinforced with the transfer of four divisions of the Potomac army's 11th and 12th corps. Another four divisions of the Army of the Tennessee marched east from Mississippi to bolster Chattanooga's defenders. With the arrival of Major General Ulysses S. Grant, appointed to overall command in the west as head of the newly created Military Division of the Mississippi, plans were implemented to lift Chattanooga's siege.

In the meantime, Thomas' army underwent a thorough reorganization and a new corps, designated the 4th, was formed. The four regiments of Harker's brigade were transferred from Wood's division to that commanded by Major General Philip H. Sheridan, and with five additional regiments constituted the 3rd Brigade, 2nd Division, 4th Corps. Harker remained in command, his brigade now composed of the 3rd Kentucky, 64th, 65th and 125th Ohio, and 22nd, 27th, 42nd, 51st and 79th Illinois.

On November 16 the Ohio "Tigers" welcomed 83 officers and men of Company I to their ranks — and to "a full share of short rations and arduous duties." Barely a week passed before the newcomers received their first taste of combat.

Head-Quarters 125th Ohio Vols.,
Before Chattanooga, Nov. 28th, 1863.

Eds. Herald — Yesterday I enclosed to you a hasty account of the three days battle in which we were engaged, attempting to give a general idea of the plan of attack.[1] I had described our advance up to the capture of Missionary Ridge. Ere that work was entirely accomplished night had spread its dark mantle around us. Our men were completely exhausted. The moment a halt was ordered, their weary limbs were stretched upon the ground. The full moon rose in queenly majesty, and the innumerable starry host adding their twinkling light made the scene one of enchanting loveliness. It was a proud moment in our soldier history. We

Opposite: Although Joseph Bruff reached field-grade rank and was in charge of the 125th at its muster-out, his personal relationships with Opdycke and several other regimental officers badly deteriorated during the war's last year. After Ohio Governor John Brough overlooked opposition to Bruff's promotion to major, Opdycke "retaliated" in September 1864. He termed Bruff "incompetent to command," as well as being "deceptive, unscrupulous and mischievous." Following arrest, charges were preferred against the officer that included "Conduct prejudicial to good order and military discipline," and "Violation of the 77th Article of War."

A month earlier the usually pious Lieutenant Colone! David Moore had become so exasperated with Bruff he wrote in frustration to Opdycke: "Would God that B. was *out, clear out, out* and *gone!* As long as I am in command of the regt., I will rule B. or kill him." Bruff weathered the harsh words. A court-martial never convened, but he faced an examining board November 10-13, 1864. In spite of Opdycke's accusations the board found his capacity "good," his qualifications "sufficient" and efficiency "fair." It was lukewarm vindication — enough to sustain Bruff in regimental command from the battle of Nashville onward and allow his promotion to lieutenant colonel in January 1865. He was breveted colonel by the War Department in 1867. Carte de visite by M. Witt Photographer, Columbus.

had driven the enemy from his stronghold. We could now look back over our old encampment on the plains below, conscious that it was forever safe. A few moments for rest — and countless camp fires blazed up on every side. Soon a hasty cup of coffee was prepared and drank, and we were again in line marching in pursuit of the foe. It was 2 o'clock in the morning before we made a final halt. Up to this time our (Granger's) corps had captured thirty cannon, more than four thousand prisoners, and several battle-flags.

We were now ordered to return to camp. During the fight Col. Opdycke was in command of Harker's old brigade.[2] His gallant conduct added much to his already high standing as a brave and noble officer.

E.P. Bates, Capt. of Co. C, commanded the regiment and behaved bravely. We lost two killed and thirty wounded. Reuben Bunnel, Co. A, and Wm. Miller, of Co. E, were killed. Capt. Bruff,[3] of Co. A, was the only officer wounded. He received a slight flesh wound in the right side. The ball passed through his pocket-book, which he carried in his side pocket, "placing," as he says, "the rebel stamp on his money," of which he had at the time over two hundred dollars.

We are still under marching orders, the men having accoutrements on, every moment expecting the signal to fall in.

CEYLON

— *Cleveland Herald,* December 8, 1863

Chattanooga, Tennessee,
Nov. 29, 1863.

Eds. Chronicle: I have delayed writing until now that I might be enabled to furnish an accurate list of the killed and wounded in the two companies from Trumbull County; also the full particulars concerning the part our regiment took in the late battle [of Missionary Ridge]. We have won a great victory, and can think nor talk of nothing else. Even the wounded soldier forgets for the moment his sufferings, while he contemplates the great work he has, with his fellow soldiers, accomplished. Let me briefly give you a general idea of the position of our forces prior to the engagement, and then refer more particularly to our regiment.

Lookout Mountain is three miles from town, next the river, and terminates in a bluff 500 feet high, and has been occupied by the

Ross County Historical Society – Chillicothe

Major General Gordon Granger, 4th Corps commander at Missionary Ridge, was flanked in this rare tintype portrait by his 2nd Division commander, Major General Philip H. Sheridan (left), and Colonel Charles G. Harker, commander of Sheridan's 3rd Brigade. Soldiers of "Little Phil's" division were among the first to reach the crest November 25. Sheridan lauded their efforts in his after-action report: "The gallant color bearers, officers and men, who planted their flags upon Mission Ridge are the true heroes of the battle."

rebels since the battle of Chickamauga. It commanded both the railroad, and the river; thus we were prevented from getting our supplies through this channel, and as the roads had become impassable, something must be done or our army would soon be in a starving condition.

Gen. Hooker[4] was on our right, below Lookout Mountain, Gen. Granger in the center and in front of Chattanooga, while Gen. Sherman, on the north side of the river, was on the left.[5] On the 23d inst., at 1:30 P.M., our center moved out one mile and drove the rebel pickets from their position.[6] On the 24th we advanced still farther, while our batteries took an active part in the engagement. At the same time Hooker stormed and gallantly carried the works on Lookout Mountain. The musketry firing was kept up until 12 o'clock at night. We have taken 2,000 prisoners, with a loss to ourselves of 250 killed and wounded. While this battle was occupying the attention of the enemy, Gen. Sherman hastily threw a pontoon bridge across the river, seven miles above, and crossed his whole force, getting in the rear of the right wing of Bragg's army.

Thus on the evening of the 24th, Gen. Hooker had taken Lookout Mountain, Gen. Hazen's brigade had captured a regiment of Alabamians,[7] while Gen. Sherman waited for the order to attack the enemy's flank; nor had he long to wait, for on the 25th day, which will ever be memorable in the history of the rebellion, the order was received to attack the enemy, but the position was too strong, and he was repulsed twice with considerable loss.

While this attack was being made, our center again advanced and took some rifle-pits near the base of Missionary Ridge. This ridge is three miles from town, and is in full view for several miles. Bragg's Headquarters was located there, directly in front of our brigade as it then stood at the foot of the Mountain in line of battle. The center was ordered to advance up that ridge, take some rifle-pits, and there rest until further orders. They obeyed the order, so far as taking the rifle-pits was concerned, but *they did not stop there for orders.* They charged right up the steep mountain side, right up to the very mouths of the rebel guns.

Standing at Fort Wood,[8] we could see the whole thing. A more grand or sublime sight has never been presented to view since the war began. The long line of soldiers, with the colors of each regiment floating to the breeze, can be seen toiling slowly, but steadily, up, up, while from forty pieces of artillery the enemy rained their shot and shell into the lines of our brave boys; still they waver not, but move on, on, towards what seems like almost certain destruction. Our batteries in our rear thundered forth their missiles at the enemy, thus forming a roof of smoke and iron hail for our soldiers, who still move on and upward; but see the line has halted; they fall back to the breastworks; Gen. Sheridan rides up and says, "That's right, boys, you done just right. You couldn't go up there alone." Then, after they had breathed a few moments, the command rings on "forward," when every man springs to his feet and again presses forward. The panic-stricken enemy throw down their arms, and while hundreds escape, many with hands uplifted rush towards our lines and gave themselves up. Col. Harker — the noble commander of our brigade — seizes a flag, and in advance of the rest, is soon astride the gun "Lady

"... about 5 P.M. our Corps stormed the
hights [sic] of Missionary ridge. It was
grand & sublime. Drove the enemy. took
many prisoners & guns. 125th did
splendidly under Capt. Bates. I had 2
horses shot under me but they were not
killed. My demi brigade took a battery ...."

• Emerson Opdycke diary, November 25, 1863

Buckner,"[9] shouting "rally round the flag boys," while one of the
soldiers of our regiment is soon beside him, and begs leave to turn
the gun on the flying enemy, which he does with good effect.[10]

Col. Opdycke exhibited his accustomed bravery, laughing at the
fears of some, or with revolver in hand threatening to shoot down
the coward who sought to skulk away to a place of safety. The
Colonel commanded a demi-brigade, and I am happy to say that
none from our regiment were threatened with his revolver. It was
Col. Opdycke's command that captured Gen. Bragg's headquarters
and eight pieces of artillery. The Colonel had two horses shot
under him, but came out of the conflict unscratched.[11] We all feel
proud of such a commander, and will follow or go at his bidding
anywhere. Capt. Bates, of Company C, commanded our regiment,
which formed a part of Col. Opdycke's command. Capt. B. is a
brave and conscientious man, and highly esteemed by all who
know him.

That night the camp fires of Lookout Mountain and Missionary
Ridge were kindled by Union soldiers, and as they gathered round
to cook their coffee, and eat their hard bread, and talked over the
scenes and incidents of the day, they grasped each other by the
hand, as friends who have long been separated. Ah, their day's
work will form one of the brightest pages in the history of our
country, and will it not hasten the time when all our enemies shall
be conquered, and we be permitted to return to home and friends
again.

The loss in our regiment was two killed and thirty wounded.[12]
All are in Hospitals and under charge of our own Surgeon. Lieut.
Moses[13] has returned to us a Captain; and Lieut. Powers is with

A medical student before enlistment, Private James Foster Scott, Company F, was the regiment's acting hospital steward when he "borrowed a gun and went to the top" of Missionary Ridge with his company. Nearly all of Scott's service was spent as a nurse or hospital clerk.

us again; we were happy to see them. They brought with them a number of men who had been absent at the various Hospitals, and at home. Our regiment is again on the march, but will probably return to this place. Adams Express is now established here, and in running order.

Yours,   H.

— *Western Reserve Chronicle,* December 16, 1863

Knoxville, Tenn.,
December 12th, 1863.

Eds. Chronicle: After three months absence from my Regiment, during which I enjoyed the delightful society of my many good friends on the Reserve, I am again returned to duty ....

Leaving home on the 16th of last month, we (Capt. Moses and his recruiting party) arrived at Bridgeport, a point on the Tennessee river forty miles below Chattanooga, about noon on Tuesday the 24th ult. All along the line of railroad extending through Ky. and Tenn. (over 230 miles) the sad havoc of war has made its impress. One is most amazed, however, by the number of troops he sees. Every town, bridge and culvert along the route is guarded by soldiers, varying in numbers from part of a company to an entire brigade.

At Bridgeport we took the boat, transport "Paint Rock," and by 4 o'clock were steaming up the river in good style. A dense fog coming on at dusk, we were obliged to tie to a tree until late next morning, before it cleared away sufficiently to allow us to proceed with safety. Several difficult places variously known as the Pot, Pan, and Suck, where the steep rocky banks hug the river closely and the waters rush along with great swiftness, delayed us considerably. At the place last named all hands were ordered ashore with cable and capstan, and worked vigorously more than an hour assisting the boat to stem the rapid current.

By noon we neared Lookout Mountain, whose rugged form commencing at the margin of the river rises to a height of 1200 feet. Only the day before it was held by rebels, and rebel batteries advantageously posted threatened quick destruction to every thing loyal coming within range. Our Captain cast many a suspicious glance, and several times brought his glass into requisition before he could be fully satisfied that all was right among the rocks, and that Lookout had indeed taken the oath of allegiance.

It was nevertheless true. At midnight of the previous day Hooker had driven the last Butternut from the mountain. With the Stars and Stripes proudly floating at the top of the jack staff, the "Paint Rock" was the first boat to pass Lookout after the fight at Chickamauga. She was hailed with loud cheers by our brave men on either bank as she passed on, and was awaited at the Chattanooga landing by scores of empty army wagons ready to convey to hungry soldiers then under fire of the enemy, the long looked for rations.

At half past one p.m. Wednesday, Nov. 25th, we passed through the streets of Chattanooga to the camp of the 125th, which we found just back of the town. Only a few sick and convalescent soldiers were there to welcome us. The regiment had been out since early on Monday, continually skirmishing by day, and resting on arms at night. The battle was now raging with great fury. We had heard the distant roar of artillery all the morning as we came up on the boat.

On going to a slight eminence back of the camp the opposing armies were in view. Circling from near the river on our left, two miles in front and terminating in Lookout Mountain on the right, Mission Ridge raised its bold front five hundred feet above the level. Long lines of our army could be plainly seen advancing near the base of the Ridge; while, from eminences to the rear our artillery was being worked to the best possible advantage. On the other hand the enemy was not inactive. The side and summit of the Ridge were wreathed in smoke from his numerous batteries. But he could no longer sufficiently depress his pieces to materially injure our infantry, who, regardless of the storm of iron and lead that raged above them, pushed steadily forward up the steep ascent and over the rebel works.

It was a glorious sight. Few men have seen an equal to its terrible grandeur — none have seen it surpassed. In the clear sunlight of that bright November day, under an arch of hissing shot and bursting shell, the brave defenders of our country's honor successfully charged the motley hosts of rebellion and drove them from one of the strongest naturally fortified places on the Continent. What wonder then that every soldiers' heart beat faster and his inspirations became more rapid as he rallied to the assistance of the dear Old Flag as it was planted in triumph over the rebel stronghold. The very boldness of the undertaking and the dauntless manner in which it was executed completely confound-

Knox County farmer Henry Whitmer suffered a severe gunshot wound in his left thigh while attempting to capture a Confederate cannon near the eastern base of Missionary Ridge. The Company F corporal returned to duty in mid-February 1865 after 14 months of recuperation at Nashville and Jeffersonville, Indiana. Carte de visite by E.W. Mealy's Photograph Gallery, New Orleans.

ed the enemy, rendering him powerless to resist the daring impetuosity of the charge. It was afterward remarked by rebel prisoners that "The men who could make such a charge could storm Hell and take the Devil by surprise."

No part of our forces was so successful as the Old Army of the Cumberland. Sherman, commanding the left, did harder fighting, but accomplished less, failing to turn the enemy's right flank as was anticipated, while Hooker, who had more difficult ground on which to fight, was confronted by very small force and sustained comparative small loss.

Having attained the summit of the ridge and captured several thousand prisoners, our wearied forces halted just at dusk to rest and prepare supper. It was indeed a meager meal; a little coffee and hard tack at best, and in many cases only a drink of cold water. At this juncture, after innumerable inquiry and much difficulty, we succeeded in joining our Regiment. The boys were very glad to see us. They forgot the hardships and dangers of the day in their eagerness to hear from relatives and friends at home. Never did we exchange more heartfelt congratulations than we did that night with our fellow soldiers on the summit of Mission Ridge. But many were missing; their places in the ranks were vacant. Three had fallen that day to rise no more, and twenty-nine had been borne from the field wounded. The bloody battle at Chickamauga had also told heavily on our noble regiment; and those who remained were dearer on account of the many dangers they had encountered.

Supper over, we were ordered to the front where we found Col. Opdycke with the remainder of his command — Harker's old brigade and one regiment, 79th, of Illinois troops — in high spirits, anxious to proceed. We pursued the retreating enemy until midnight when we bivouacked for the remainder of the night on the banks of the Chickamauga, being unable to proceed further on account of bridges being destroyed.

Next day we returned to camp, our Brigade having captured during the engagement eight pieces of artillery and one thousand prisoners.

Although no official statement has yet been made, the entire loss of the enemy will probably not fall short of 12000 in killed, wounded and prisoners, 80 pieces of artillery and innumerable small arms.[14] Our own losses will scarcely be one third as great. Besides, the advantage gained to us is beyond estimate. Had we been unsuccessful, Knoxville would to-day have been in the hands of Lee, and another year would have been added to the rebellion.

Had I not already protracted this letter to undue length, I would speak of the personal bravery of many officers and men, and relate interesting incidents of the battle. Col. Opdycke handled his command splendidly, displaying great coolness and bravery under every circumstance. He did not detract from the high reputation he attained at Chickamauga for ability and gallantry; on the contrary, he gave new proof of quick perception and power of execution in trying times when such qualities were most necessary. Capt. Bates commanding the regiment, and Lieut. S.A. Smith, acting Adjutant, did nobly, winning high praise for their good conduct.

In my next, which may possibly accompany this, I will give you an account of our march into East Tennessee.

Respectfully,   CEYLON

— *Western Reserve Chronicle,* January 6, 1864

Knoxville, Tenn.,
Dec. 14th, 1863.

Eds. Chronicle: Returning from the battle of Mission Ridge we rested one day in camp when we received orders to prepare for marching. Knoxville was the place of destination, and as only one wagon was to be allowed to each regiment, each officer and man must carry blankets and rations to make himself comfortable.

At 2 o'clock P.M. Saturday, November 28, leaving all the sick, wounded and those least able to march behind, we took a farewell look at our comfortable quarters at Chattanooga, and took line of march. We left the river to our left, and plodding through mud and rain, halted for the night in a dense woods, having made only seven miles during the afternoon. Next day the weather brightened. By dark we found ourselves within one mile of Harrison's Landing; meantime having been considerably delayed by swollen streams. We bivouacked on a beautiful side hill. The flames from our thousand camp-fires beautified the night, and shone in pleasing contrast with the myriad twinkling stars that gemmed the blue arched Heavens. Cattle, which we had driven with us, or captured on the route, were killed, and a two thirds ration of meat issued to the men. After supper, an order came from headquarters that Burnside was closely invested at Knoxville, with only six days' rations for men and animals, and we must make that point, 120 miles distant, within the time.[15]

Every man responded cheerfully that he was ready and anxious to make the attempt. Next morning we were in line ready to advance by three o'clock. By two o'clock P.M. we had marched twenty miles and commenced crossing the Hiwassee river. As our advance approached the ferry, a squad of rebels fired from the opposite bank without effect, and hastily fled. The remaining part of the afternoon and the entire night were occupied in crossing our brigade with artillery and wagons. Col. Harker was indefatigable in his exertions to facilitate the undertaking, laboring constantly at the oar until it was accomplished.

I will not further weary your patience by describing each step of our army as it advanced. Remembering the old adage, "Large bodies move slow," you will not be surprised to learn that we did not reach Knoxville by the expiration of six days. We did, however, accomplish the object for which we set out. Longstreet taking fright at our approach, made one desperate attempt to take the city by assault, and retired mad and disheartened at the barren result of his daring experiment. That charge cost him one thousand of his best officers and men.[16] Poor fellow! He will know better next time how to estimate the valor of Western troops.

The following are among the list of troops that left Chattanooga for the relief of Burnside: Granger's, Howard's, and a part of Sherman's command.[17]

Quite a number of the above returned before reaching this place. Our own, Granger's corps, encamped opposite the city Monday, Dec. 7th, and on the morning of the 9th, our brigade crossed the Holston River and went into camp one mile back of town in the immediate vicinity of Longstreet's fortifications. Our men were much fatigued, and many of them nearly shoeless. We had all the time been on half or two-thirds rations, depending on the country through which we marched, for supplies.

Knoxville, as do also Chattanooga, Tullahoma, Murfreesboro, Franklin, and all the Southern towns which we have visited, gives unmistakable evidence that the blighting hand of war has been upon her. Over one hundred houses were destroyed by Burnside, in order to open range for his artillery during the siege, and prevent them from being occupied by rebel sharp-shooters. Fences for two miles in every direction have been consumed for fuel, and all the public and many of the private edifices in the place are now occupied as Federal Hospitals, or store-houses for army supplies.

Provisions are held at siege rates, as the following current price list will show: Butter, $1 to $1.50 per lb; eggs, 50 cts. per dozen; chickens, 50 cents to $1 each; turkeys, $2 and scarce; coffee, $2 per lb; molasses, $2.50 per gallon; soda, $1 per lb; soap, none in the market; cider, $25 per bbl; apples, 25 cents per dozen; sugar, 75 cents per lb; boots, $15 per pair.

Since writing the above, I have learned through a courier that our cavalry are still closely pursuing Lee, who is making for Virginia with all possible haste.[18] Yesterday Shackleford captured three hundred of his rear guard, and stragglers.[19]

Lee is proving himself to be a poor itinerant General, and if he does not have a turn of luck soon, he may have occasion to console himself with the old song, "Ha'nt I a used up man."

CEYLON

— *Western Reserve Chronicle,* January 6, 1864

Notes to Chapter Three

1. A letter from "Ceylon" dated November 27 did not appear in either the *Cleveland Herald* or *Western Reserve Chronicle.*

2. From November 23 to 26, Opdycke commanded a demi-brigade of five regiments — 3rd Kentucky, 79th Illinois, 64th, 65th and 125th Ohio. *OR,* vol. XXXI, pt. 2, p. 233.

3. Born in Mahoning County, Joseph Bruff was a farmer residing at Damascusville, Columbiana County, prior to his August 1862 appointment by Governor David Tod as captain of Company A. Joseph Bruff CSR, RG 94, NARA.

4. Major General Joseph Hooker commanded the 11th and 12th corps. For the November 24 operations at Lookout Mountain he had at his disposal three divisions, one each from the 12th, 15th and 4th corps. The 11th Corps had been sent to Chattanooga November 22. *OR,* vol. XXXI, pt. 2, p. 314.

5. Major General William T. Sherman had immediate command of four divisions at Chattanooga, three from his Army of the Tennessee and one from the 14th Corps. *OR,* vol. XXXI, pt. 2, p. 572.

6. Orchard Knob, a low ridge halfway between Chattanooga and Missionary Ridge, had been occupied by Confederate picket outposts. Boatner, p. 144.

7. Hazen reported capturing 146 officers and men on Orchard Knob. *OR,* vol. XXXI, pt. 2, p. 280. The prisoners belonged to the 24th and 28th Alabama regiments.

8. Fort Wood, named for Union General Thomas J. Wood and one of the strongest points along Chattanooga's defensive lines, was renamed Fort Creighton in 1864 to honor Colonel William R. Creighton, 7th Ohio, who was killed in action November 27, 1863 near Ringgold, Georgia.

9. "Lady Buckner" was a pet name given to one of the guns comprising Cobb's Kentucky Battery (C.S.) in honor of Major General Simon B. Buckner's wife.

10. The 125th's regimental history identifies this man as Corporal Reuben M. Steele of Company I. Clark, p. 174.

11. Partway up Missionary Ridge Opdycke's favorite mount "Barney" was hit in the mouth by a bullet, which cut the bridle and made the horse unmanageable. First Lieutenant Abner B. Carter, the 125th's quartermaster and Opdycke's acting aide, gave the colonel his own horse, and that animal was wounded in the shoulder near the ridge's crest. Opdycke proceeded on foot to the summit. There a member of the 65th Ohio shot a mounted Confederate from his saddle, and Opdycke appropriated the animal for the rest of the battle. *To Battle for God and the Right,* p. 135-136.

12. Four of the injured subsequently died of their wounds. Clark, p. 178.

13. Elmer Moses of Hartford, Trumbull County, previously served under Opdycke as a sergeant in Company A, 41st Ohio. He had been first lieutenant of Company B, 125th Ohio.

14. On December 1 General Thomas reported "at least 225 rebels killed"

in the battle of Chattanooga, and nearly 5,500 Confederates captured during all of November's operations. Matériel captured included 40 cannons of various types and 6,175 stand of small arms, mostly Enfield rifle muskets. *OR,* vol. XXXI, pt. 2, p. 98-100.

15. Early in November, Longstreet and Confederate troops under his command were sent from the Chattanooga area to attack Major General Ambrose E. Burnside's Union forces at Knoxville, Tennessee. Unable to undertake a regular siege of the town, Longstreet, after receiving additional reinforcements, assaulted Fort Sanders at Knoxville November 29, but was repulsed. Boatner, p. 467-468.

16. Confederate losses at Fort Sanders were reported as 129 killed, 458 wounded and 226 missing. *OR,* vol. XXXI, pt. 1, p. 475.

17. Brigadier General Jefferson C. Davis' division of the 14th Corps and a small body of cavalry also accompanied the expedition to Knoxville. Clark, p. 185.

18. "Ceylon" was in error. General Robert E. Lee was not in East Tennessee. Longstreet, Lee's subordinate, retired from Knoxville to Greenville, Tennessee, and went into winter quarters. Boatner, p. 468.

19. At the time, Brigadier General James M. Shackelford commanded the Department of the Ohio's Cavalry Corps. His December 13 report does not mention the capture of 300 Confederates, as recorded by "Ceylon." *OR,* vol. XXXI, pt. 1, p. 414-415.

1864

FOUR

'One of our fiercest tiger yells had the desired effect'

The new year of 1864 opened rather gloomily for the 125th
Ohio. Harsh East Tennessee weather, insufficient clothing,
food, supplies and scanty shelter combined to make life "discouraging" in the regiment's ranks during much of their winter sojourn in the rough countryside east of Knoxville. A bright moment occurred January 14 when Lieutenant Colonel David Moore
arrived at Blain's Crossroads, bringing with him newly mustered
Company K and a sizeable number of convalescents. The regiment finally consisted of 10 companies, and Moore took command.
Just three days later his leadership underwent its first test when
the 125th tangled with a portion of General Longstreet's Confederate forces at Dandridge, Tennessee. In his diary, Opdycke labeled the affair "a severe skirmish," and proudly noted the 125th
"behaved gallantly" and "fought like Tigers."

Camp near Clinch Mt., Tenn.,
January 6th, 1864.

My Dear Grandfather I have been waiting a long time for a
good opportunity to write to you. I had hoped to find myself seated, with pen, ink, stand and comfortable quarters, where I might
tell of the many things interesting to you, which have happened
since I last wrote, but that time is not yet. Making virtue of necessity, I cheerfully accommodate myself to circumstances beyond
my control, and write with pencil (your specks will come in good
play), surrounded with forest trees, and a troop of shivering comrades, all eager to get a chance at the large log heaps blazing before me.

Since leaving Chattanooga, we have been entirely without
tents; nor have we stayed long enough in any one place to war-

rant us in erecting log huts, or other suitable protection against the weather. We have suffered some from cold, as many a burnt blanket, scorched coat or trowsers will testify. Not a few army shoes, and even boots of Trumbull manufacture, bear marks of having been placed in too close proximity to the fire. One of our boys had his overcoat half burnt off him the other night; and Capt. Bates slept soundly while my rubber, and another blanket, were almost entirely consumed at his side.

We have occasionally felt some inconvenience from hunger. On account of deficient transportation, we have seldom had more than half rations, and some of the time have been unable to procure forage from the country. After all, we have been highly favored; and, I do not think one could be thought superstitious for seeing the hand of Providence kindly turning many circumstances to our advantage. We have as yet had no snow, Winter thus being robbed of half its rigor. Neither have we had heavy rains as might have been expected at this season of the year; notwithstanding, there has been a good stage of water highly favorable to the transportation of supplies by river.

You desired me to tell you about the people and country. I can say much in favor of both. Beautiful mountains lifting their rocky peaks heavenward meet one on almost every side. Here they stand boldly forth. You can count their rocky ribs, as cliff on cliff is piled before you. Yonder they gradually recede toward the setting sun until their blue tops blend with the skies in the horizon. The country bears many marks of genuine enterprise. The villages in many places are under a high state of cultivation, and neatly enclosed fields and comfortable looking houses are frequently met with. The people are nearly all Union. They manifest a purity of patriotism and depth of devotion that should bring the blush to the face of Northern Copperheads. Their love for the old flag has been purified through suffering; and their firm loyalty unshaken by the wild storm of treason that has so long raged about them, spreading devastation and terror on every side, is above reproach, challenging the admiration of the world. "I would rather that every cent I possess should go toward subsisting your men, than allow the rebels again to hold this country," is an expression I have several times heard repeated. East Tennessee is not yet relieved from suffering. As Longstreet passed through, his men robbed and plundered everything. They declared it their object to desolate the country so our army would be unable to pursue them. As a consequence, the people are without provisions. Many must depend upon the Government, in part, for subsistance, while others will move North until a better state of things can be established.

There is not much prospect of further active operations during the winter. We may, perhaps, make one more attempt to engage Longstreet before going into permanent quarters.

Many of the old regiments are enlisting as veterans.[1] Fully one half the soldiers from Ohio have probably re-enlisted by this time. Our regiment is too young to receive the benefit of the older; though if deeds were taken into consideration, we might find ourselves in advance of many older organizations. I think Home Guards should be ashamed of themselves, loitering behind, while brothers and friends are enlisting for another three year term in a cause dear alike to all, and worthy the best blood of the nation.

The health of our regiment was never better than now. Many of those wounded in the battles of Chickamauga and Mission Ridge have again returned to duty. We number nearly four hundred effective men ....

I am your obedient grand-son.
CEYLON
— *Western Reserve Chronicle,* January 27, 1864

Camp 125th O.V.,
January 21, 1864.

Eds. Herald — We have again demonstrated to Mr. Longstreet that he cannot whip us. Our men possess courage that superior numbers cannot overawe. Let him who doubts it look at Chickamauga, Mission Ridge, and this last uneven contest at Dandridge.

For several weeks past it has been in contemplation to send the troops in East Tennessee into winter quarters. Much difficulty has been experienced to find a suitable place. We must hold Longstreet in check, at the same time depending largely upon the country for supplies. The long line via Chattanooga, Nashville and Louisville, three hundred and fifty miles, guarded by Federal bayonets, is inadequate to our demands. It would be equally impossible, considering the wretched state of the roads, and the emaciated condition of our mules, to draw supplies across the country through Cumberland Gap. The country anywhere within twenty miles of Knoxville has long since been exhausted. Hence, there is no alternative; we must depend upon the country farther away for supplies, or withdraw a part or all of our forces from East Tennessee.

On Thursday of last week we commenced leaving Blain's Cross Roads with the intention of going into winter quarters on French Broad River. Our (Harker's) brigade broke up camp on Friday. On Friday we bivouacked near Dandridge, 34 miles east of Knoxville. There had been some skirmishing during the day between cavalry, but the rebels did not seem to be in force, and retired during the night. Next morning our regiment was ordered on picket one mile east of the town, supposing that, as a matter of course, the remainder of the brigade would be held in reserve. The 93d Ohio was posted to our right, while our left was badly exposed, only a few cavalry videttes having been thrown out in that direction.

At 2 o'clock p.m. the enemy made his appearance, firing occasional shots at our cavalry videttes. All the fore part of the day the rebel pickets and ours had been in plain sight of each other, and at the commencement of the firing it was thought to be only an exercise for the amusement of the pickets. We were, however, soon undeceived. From a commanding eminence to our left a long column of rebel cavalry could be plainly seen descending a hill and forming in line of battle in a wooded valley to our front. Their skirmishers soon detected the exposed condition of our left, and were not slow in taking advantage of it. A very heavy skirmish line of dismounted cavalrymen came in on double quick on our left. The cavalry in front of us retired after a few shots, and before we could change our line they were upon us. We doubled our skirmish line from our advanced reserves. In changing front our left was forced back nearly upon the main reserve.

The rebs then tried to force back the right with the seeming intention of surrounding and capturing the entire regiment. But their charge was bravely met. Taking advantage of the ground, we twice received their charge and both times repulsed them handsomely. They then opened upon us with two pieces of artillery, but put in too much powder and shot over every time. They then charged for the third time upon the right, driving the left back upon the grand reserve and the right to a line almost even with it. The grand reserve opened, firing volleys by rank. This was too much for the graybacks; they came to a dead halt. Just at this time companies A and C, thrown out as skirmishers to the right, charged with a tiger yell, and drove them in confusion over the ground occupied by our pickets in the morning, to their second line. It has been reported by cavalrymen posted on an eminence to our right, that the second line gave way and fell back half a mile before they were again rallied. Night now began to close in around us, and our men were too much exhausted to follow up their advantage or even sustain another charge should they again bring fresh troops against us.

Lieut. Colonel Moore, commanding the regiment, having learned that our brigade had moved early in the day, and that there was no other infantry within supporting distance, was conscious of the danger to which he was exposed and improved his first opportunity to find relief. Unable to find an officer superior in rank, he reported to Col. Garrard[2] of the 7th Ohio Cavalry, and by hard urging induced him to send his own and the 2d Michigan cavalry to relieve the 125th. We then retired over half a mile toward town and then were posted near the road on a wooded hill.

It was about an hour before the rebs again rallied and came up to the work. When they did, they were met with great bravery by our dismounted cavalry, who, however, were unable to hold the ground against such heavy odds. They were driven back about a quarter of a mile when the rebs were again repulsed and did not again renew the attack.

The following is the loss of the 125th during the day: Adjutant Smith, killed. Lieut. Clark of Co. H was hit with a spent ball, but not seriously injured.

Privates killed — Richard P. Likens, Co. D; Conrad Ling, Co. F; George Beckwith, Co. H.[3]

Privates wounded — Mathias Callahan, Co. A; John Boner, Co. A, severe in side; F.J. Fobes, Co. B, slight in left leg; John D. Mahan, Co. C, severe through right lung; James B. Scott, Co. E, slight in back; Henry Graham, Co. F, severe in left side and arm; Cassius C. Burch, Co. H, slight; A.J. Couch, Co. I, slight in breast.[4]

Lieut. Smith was shot in the breast by a musket ball, and expired shortly afterwards. To within a few months he had been Sergt. Major, and on all occasions proved himself to be a brave and efficient officer.[5]

Having lighted many fires along the side of the hill to give it the appearance of a large encampment, we commenced to retreat about 9 o'clock at night, having been preceded by a large part of our force during the day. It was to prevent bringing on a general engagement that the 125th and the 93d Ohio were not reinforced;

and, although we were not pleased with the arrangement at the time, we are better satisfied it should be so than that we should have suffered severe loss.

We had not ammunition to fight a battle, nor was that any part of our object in going so near to Mr. Longstreet. "He who fights and runs away, may live to fight another day," is appropriate in our case; and if Longstreet thinks fight is not in us, we invite him again to Knoxville, where we will dictate the kind of weapon, and he may determine the distance in order to settle the question.

We are now within 15 miles of Knoxville. We are satisfied that we have inflicted as heavy loss on the enemy as ourselves have sustained. Our army is on less than half rations, but our pluck never was better. We have plenty of warm clothing and brave hearts to keep us warm and cheerful. The more we suffer, and the more of our brave comrades we see fall about us, the more we are determined to be avenged, and the more we love the cause in which we are engaged.

Old regiments have nearly all re-enlisted. In less than two hours after the rolls were presented nearly every man in the 125th had signed them. As we have not been in the service long enough to come under the order, application has been made, warmly endorsed by the Generals under whom we have fought, asking for a special order admitting us to the privileges of veteran troops. Whether admitted or not, depend upon it we are true to home and country, and will not fail to stab treason at every vulnerable point.

CEYLON

— *Cleveland Herald,* February 3, 1864

Camp near Strawberry Plains, Tenn.,
January 22, 1864.

Eds. Chronicle: During the latter part of last week our forces left their encampment at Blain's Cross Road, marching to the front. On Saturday, the 16th inst., they concentrated at Dandridge, a town on French Broad River, thirty-four miles east of Knoxville. On the evening of that day a sharp skirmish between our advance and Longstreet's Cavalry took place a short distance east of the town. The rebels retired during the night, but appeared again early next morning and established their pickets in plain view of our own.

L. M. Strayer Collection

Charles T. Clark, the 125th Ohio's youngest commissioned officer, served at various times with Companies F, H and G. Promoted to captain late in 1864, he was detached for brigade staff duty during the summer of 1865. Thirty years later his regimental history *Opdycke Tigers 125th O.V.I.* was published in Columbus.

The 125th occupied the road most threatened, while the road to our right was held by the 93d Ohio. Our left was badly exposed, being protected by only a few cavalry videttes. Firing between the outposts commenced at 2 o'clock P.M. and it soon became evident that an attack was meditated. Long lines of rebel cavalry could now be seen descending a hill to our front and forming in line in the ravine below. Dismounting, they first felt of the line to our right, but drawing a sharp fire from the 93d, they soon circled to the left, and came in with a shout on our unprotected left flank. The Enfield's fire was too much for our men, who were forced to fall back nearly to the guard reserve. An attempt was then made to force back companies A and C deployed as skirmishers to the right, but, aided by a company from the 93d Ohio, they repulsed two desperate charges and only retired slowly when charged for the third time by more than three times their number.

Falling back on a line even with the grand reserve, it became evident that to assemble on the reserve, or retire further, would let the rebels in our rear and expose the entire command to capture. We made a vigorous charge, giving as "Artemus Ward [6] would say," one of "our fiercest tiger yells." It had the desired effect. The rebels ran like scared sheep. We drove them past the point occupied by our advanced reserves in the morning and only ceased pursuit after drawing fire from their second line. Cavalry posted on an eminence to our right say the second line broke and retired half a mile before it was reformed.

Night now coming on, our exhausted men assembled on the reserve, and Lt. Col. Moore, commanding Regiment, conscious that another charge made by fresh troops might prove disastrous, applied to be relieved.

The remainder of our brigade had retired early in the day, and no other force could be found near except the 7th Ohio and 2d Michigan Cavalry regiments. After hard urging they dismounted and came to our relief. We then fell back nearly to town, resting on a wooded hill which we soon illuminated by carrying rails and building numerous fires.

We now began to enquire for our dead and wounded. Among the former was Lt. Smith, acting Adjutant, shot through the head. Lt. Smith had been our Sergt. Major since the organization of the regiment, to within a few months past, and was highly esteemed by us all. Uncoffined, we buried him in the Dandridge Cemetery. A genial companion and a brave christian officer; his illumined

When Second Lieutenant Seabury A. Smith was killed at Dandridge, he had been an officer just three months, his promotion from sergeant major coming while recruiting for the regiment in Ohio. The fatal bullet, Colonel Opdycke wrote, "passed through the lower back brain when he fell from his horse and expired with a pleasant smile upon his countenance."

Clark, *Opdycke Tigers*

spirit now whispers from the heavens showering benedictions on the cause, in the defence of which he so nobly offered up his life.

The entire loss of the regiment was four killed, ten wounded, and five missing.[7] Riley Fitch and J.P. Gartner of Co. B had their guns shot from their hands; and Sergt. John Canon and Nathan Warden, of Co. C, narrowly escaped being killed by the bursting of a shell.

Our regiment was among the last infantry to retire, leaving Dandridge at 9 o'clock at night. We marched until three o'clock next morning, and bivouacked at the side of the road until daylight. At noon we crossed the Holston river at Strawberry Plains, and shortly after came into camp at this place.

Our forces are mostly encamped between this and Knoxville. It is thought Longstreet will follow and give battle.

We are on less than half rations, but our men are unusually healthy and in good spirits.

Longstreet has conscripted all the white men between the ages of eighteen and fifty, and negroes between fifteen and sixty, within his reach. He has probably thirty thousand troops.

We have here the 4th, 9th and 23d Army Corps. If properly

supplied, we can whip Longstreet on ground of his own choosing. In any event Knoxville is safe; and when our veterans return we will clear East Tennessee of every vestige of treason.

Respectfully,
CEYLON

— *Western Reserve Chronicle,* February 3, 1864

Strawberry Plains,
Jan. 19, 1864.

Dearest Wife: Have an opportunity to send you a line by Dr. McHenry, who goes with our sick and wounded to Knoxville. Last Saturday, the 16th, we reached Dandridge. There had been skirmishing all day. We were immediately ordered out to repel an attack. Colonel Opdycke's demi-brigade was held as a reserve. The rebs were driven with ease. Next morning I received an order to report with the regiment for picket duty, and took out my regiment; 93rd Ohio on my right, and a brigade of cavalry commanded by Colonel LaGrange,[8] 1st Wisconsin Cavalry, Pearly Newton's Colonel, on my left. We had a large part of our outpost line in an open field, through which the main road to the enemy passed. My reserves were posted to the right and left of this open ground in woods; my grand reserve, under my immediate control, in rear of the above. I asked in vain for another regiment to picket in rear of cavalry on the left. It was promised but it did not come.

The rebels were in plain sight. At noon the videttes were engaged; by 3 o'clock my outposts were attacked. Their reserve went immediately to their support. The fight became general and severe. The cavalry were driven back, which enabled the enemy to turn my left flank. The skirmishers then were thrown back on the grand reserve, which was now attacked vigorously. I had placed it in a semi-circular natural rifle pit, and had the men to

Opposite: Private Jefferson Melick, Company F, was one of seven "Tigers" captured at Dandridge and originally confined in Richmond, Virginia. All except Melick, who was hospitalized for "diarrhea & debility," were transferred to Andersonville prison in Georgia, where five of them died in the summer of 1864. Melick, a Mount Vernon farmer, was exchanged and returned to duty April 4, 1865. Carte de visite by Lilienthal Gallery, New Orleans.

lie down. When we were attacked I opened fire upon them by rank. I soon drove them back and silenced their fire on my left and immediate front. As soon as I would cease firing they would begin again, and then we would go at it.

The 7th Ohio Cavalry at the beginning of the fight were drawn up on the hill beyond a creek in my rear. Saw "Met" Mitchell[9] and Si. Long. Sent my compliments to Major Norton,[10] and told Mitchell to tell them, "The 125th Ohio is in your front." My skirmishers on the right, Companies A and C, under the general supervision of Captain Bates, fought magnificently. They charged repeatedly upon the enemy, and drove him back, only in turn to be driven by overwhelming numbers themselves.[11]

The enemy's artillery played upon us furiously. No support came to our rescue. The cavalry had disappeared from the hill in our rear. Our two guns there did the best they could, but could not check the enemy. Lieut. S.A. Smith, acting Adjutant, was shot dead from his horse by my side. You remember him. Another was killed on my right and one wounded on my left. Nothing but the nature of the ground saved my men from slaughter. God directed me to the position I took up. The volleys from the concealed force dismayed the enemy. Their sharpshooters fired upon us from the tree tops. Smith was killed by one. When he fell I was without an aide, and asked for a volunteer orderly to mount his horse. Little Johnny Simpson, Company G, volunteered and behaved nobly. I never saw men stick together and fight so desperately. The new company behaved like veterans. But all in vain.

It was night. No instructions, no orders, no reinforcements; the enemy in still augmenting force closing upon us, stretching away beyond both my flanks. The 93rd Ohio had retired; the 125th was alone; such being the case, I reluctantly retired my men amidst a storm of lead, across the creek up to the hill beyond, on the crest of which I re-formed them and sent in another volley, which checked their advance for the night. Several of my command were wounded whilst crossing the stream, and one killed. The orderly was shot in the arm, and at the top of the hill received another shot in the side, and his horse was killed. I shall believe that the desperate gallantry of my command saved the town and forces from capture. Still waited impatiently for orders. My flanks were unprotected, my command liable to capture.

At this juncture dismounted cavalry, under Colonel Garrard, 7th O.V.C., advanced as skirmishers to [the] crest of [the] hill. I reported to him, and received permission to return to [the] crest of [the] hill next in rear of [the] cavalry, when I had arms stacked, and after throwing out [an] advanced guard had my men rest. Could find no infantry yet, a strange thing it seemed when I considered there were two entire divisions, Wood's and Sheridan's, there that morning; neither did I receive any orders.

Sent fatigue party out on my right to build fires on the hill, to represent camps of regiments, to make the rebels believe fresh troops had arrived. Send Captain Bates, acting Major, to whose valuable services I am greatly indebted, to endeavor to communicate with headquarters for orders. He returned at length, and after a while a Wisconsin regiment of cavalry relieved us, and we, as ordered, proceeded to join our brigade, and then we learned that preparations for retreat had been going on all day. Colonel Opdycke's demi-brigade had been out hard at work building bridges. *We were left unprotected for fear of bringing on a general engagement,* as Longstreet was regarded as too strong for us. The new route was found impracticable, and the retreat to Strawberry Plains was made principally over the road upon which we advanced.

My loss is five killed and thirteen wounded, and perhaps a few missing. Colonel Harker, our brigade commander, and General Sheridan, our division commander, who witnessed part of our fighting, pronounced the conduct of the regiment magnificent ....

And so, my dearest wife, I celebrated your birthday.[12] God spared me to you and Hasie and Pa and Ma. A bullet hole through the sleeve of my overcoat shows how near to harm I came. I rode a beautiful mule, the finest I ever saw, my secesh mare having failed. At first I feared he would be unmanageable, but he soon became perfectly used to the whistle of balls. I send you the enclosed order for a birthday present.

Yours   D.
[David H. Moore]

— Clark, *Opdycke Tigers*

Notes to Chapter Four

1. The Veteran Volunteer Act, passed early in 1864, provided that any man who reenlisted for the war's duration, having two or more years' previous service, would receive a 30-day furlough, free transportation home and a $400 bounty. Boatner, p. 870.
2. Colonel Israel Garrard.
3. Privates Likens and Beckwith were mortally wounded, dying later on January 17. Private Ling, a Mexican War veteran, was shot through the head and instantly killed. Clark, p. 210-211.
4. Also wounded at Dandridge were Corporal John Simpson, Company G, and Private Walter Cheney, Company C. Only 13 years old when he enlisted in September 1862, Cheney was hit in the leg while helping to carry Private George Beckwith to the rear. Clark, p. 211, 212.
5. In his diary, Opdycke stated that Lieutenant Smith was shot "through the lower back & brain." He had been commissioned second lieutenant of Company I on September 24, 1863. *Ohio Roster,* vol. VIII, p. 443.
6. "Artemus Ward" was the pen name of popular humorist Charles Farrar Browne (1834-1867). Before the war he worked as a *Cleveland Plain Dealer* reporter. *Funk & Wagnalls New Encyclopedia,* vol. 27, p. 148.
7. The regimental history lists 125th Ohio casualties at Dandridge as four killed or mortally wounded, 14 wounded and seven captured. Clark, p. 203.
8. Colonel Oscar H. La Grange commanded the 2nd Brigade, 1st Cavalry Division, Army of the Cumberland.
9. At the time, Charles D. Mitchell was the 7th Ohio Cavalry's sergeant major. *Ohio Roster,* vol. XI, p. 369.
10. Major Augustus Norton, 7th Ohio Cavalry, resigned two weeks after the battle of Dandridge. *Ohio Roster,* vol. XI, p. 369.
11. According to Longstreet, Confederate forces engaged at Dandridge were "a part of Hood's division" and cavalry belonging to Major General William T. Martin's division. *OR,* vol. XXXII, pt. 1, p. 93.
12. Moore's wife, Julia, resided in Athens, Ohio.

FIVE

' Reading sermons in stones
and sending our prayers on bullet wings '

A week after the battle of Dandridge the 125th Ohio and its
brigade relocated to Loudon, Tennessee, where they established winter quarters and spent the next three months "in comparative comfort." In late February an army-wide inspection
found Harker's brigade "deserving especial notice; and of [all] the
regiments in the brigade, the 125th Ohio deserved notice for its
drill and cleanliness of arms." It was the only regiment in three
different corps to be mentioned and commended by name.

Opdycke was not present for the inspection, having been
granted 60-day leave on January 27 to recruit in Ohio. At Warren
he was presented by friends with a $200 watch and chain. When
he returned to Loudon April 4 he was warmly greeted by cheers
from the regiment, standing in formation. "I feel good to meet my
braves," the colonel jotted in his diary that evening.

Opdycke learned that important changes had occurred during
his absence. In March, General Grant was assigned to overall
command of the United States armies, and was succeeded as head
of the Military Division of the Mississippi by General William T.
Sherman. General Gordon Granger was relieved of 4th Corps
command and replaced by General Oliver O. Howard. Phil Sheridan, the 125th's division commander, was sent east to take over
the Army of the Potomac's Cavalry Corps. In his stead was named
Major General John Newton, a veteran of Fredericksburg, Chancellorsville and Gettysburg.

The Confederate Army of Tennessee, encamped in northwest
Georgia around Dalton, was commanded by General Joseph E.
Johnston, who had replaced Braxton Bragg in December 1863.
Destroying Johnston's army was Sherman's objective, as Grant
ordered him "to break it up, and to get into the interior of the en-
emy's country as far as you can, inflicting all the damage you can

against their war resources ...." Since Atlanta was a vital Confederate supply, communications and manufacturing center, Sherman initiated an advance from Chattanooga toward that city early in May with some 99,000 troops organized in three armies.

The 125th Ohio began the journey to Atlanta with 517 officers and men. When the campaign concluded four months later, half of them had been killed or wounded.

Nashville, Tenn., May 8th, 1864.

Eds. Chronicle: After an absence of about six months from my Regt. (125 O.V.I.), the most of the time unfit for duty, being afflicted with chronic diarrhoea, that of all diseases most fatal to the soldier.

I left Warren on the 29th of April for the front. After a pleasant trip as far as Cleveland, I left for the Land of hard tack, wondering who was responsible for all the partings of friends, sufferings of our Brave Boys in the Army, sorrows of thousands of disconsolate widows, mothers, and sisters and trials of tens of thousands of hapless orphans, until I found myself playing the part of the woman; yet I felt it no disgrace. Christ wept over the grave of a single fiend; the Servant is not above his Lord.

I do not think of any occurrence on the way worth noting; my transition though short was a little chequered. Having drawn no pay from [the] government for ten months and having a family to provide for, necessarily left me with but little money ($32 in all for which I sold an old cow and calf), but with this I started; while at home I rec'd a Lieutenant's commission.[1] Our army regulations provide that Officers shall wear a uniform according to rank, so at Cleveland I pitched in. After buying a Coat, Belt, and a pair of Shoulder Straps (an old sword I had), I found myself with a gay uniform and $12 in my pocket; with this I broke for Dixie to join my regt., which was then in front of Chattanooga.

I reported at Columbus; hoped that I should receive commutation for rations, but nary a commute. So on I trundled to Louisville, Ky. where I lost my valise with its contents, including my Belt, into which I had stowed it to save it from being soiled. The Belt being a necessary article in my outfit, there was no way left for me but to buy another, which I did in Nashville, and found by inventory, after paying other expenses, that my cash account had grown alarmingly short; but what probed deeper the wound was the loss of a case of medicines, a lot of dried fruit, and other articles carefully arranged and put up by an anxious wife and children.

While waiting for transportation at Nashville I called at the Soldiers Home and asked the Superintendent if I could stay there until I could get transportation to Chattanooga. He looked up at those straps I told you about buying in Cleveland, and remarked that the Home was for privates and non commish, so dear faithful sisters be not weary in well doing; while you aid us with your hands, give us an unbroken current of your sympathy, and we will fight for you until He who knows the end from the beginning shall say "peace be still."

Now dear *Chronicle,* I expect to join my devoted old Regt. today, and ere this reaches you we shall be trying titles with Jo. Johnston; but with that model of a commander, Col. Opdycke, the 125th has but little to fear. With the prayers of friends and the blessing of God we hope to survive the trial. Our Regt. is in splendid health, kept so by the skill and untiring energy of Surgeon McHenry.

Rollin D. Barnes
Lieut. 125 Regt. O.V.I.

— *Western Reserve Chronicle,* May 25, 1864

Headquarters 4th Army Corps,
Signal Department,
Catoosa Springs, May 5th, 1864.

Eds. Chronicle: On Tuesday, the 3d inst., at an early hour, the grand Army of the Cumberland, numbering —— men,[2] was set in motion. The 4th Corps had been stationed at Cleveland, Tennessee, for a few weeks, or rather, Gen. Howard's Headquarters were there, while the troops were scattered along the R.R. on either side of the town. This was truly a beautiful place, and the majority of the inhabitants were Union people. For some reason, the scourge of war had not produced such a desolating effect here, as elsewhere. True, business was suspended, and nearly all the men and boys were in the army; but the natural beauty of the place remained the same. The roads are in excellent condition, and the army seems in fine spirits, and confident of success.

There appears to be nothing wanting to make this campaign a

Major General
George H. Thomas,
revered by his men
and fondly nicknamed
"Old Pap," com-
manded the Army
of the Cumberland
during 1864's Atlanta
campaign.
This portrait was
made in Nashville
by T.F. Saltsman.

Courtesy of Janis Pahnke

decisive one. The baggage of officers has been cut down to less than half what it was one year ago. The six mule wagons which cumbered the roads for miles last year, and could almost be numbered by thousands, have now been reduced to hundreds. A regiment which eighteen months ago had twelve wagons and three ambulances, with seventy-two mules and six horses, now has three wagons and eighteen mules and an ambulance. True, we have a supply train, ammunition train and ambulance train, but our transportation has been reduced to a more perfect system than ever before. The soldiers have learned to do with less luggage; the officers take nothing but light valises, leaving what is not actually needed behind until the campaign is ended. Thus the army moves rapidly, and you may expect to hear that it will strike harder, oftener, and more together than ever before.[3]

Gen. Howard and staff, in advance of the Corps, reached the enemy's pickets at 12 M. yesterday, one mile beyond Catoosa Springs. The signal party always accompanies the General, and during a campaign composes a part of his staff and body guard. A light reconnaissance was made which resulted in driving in the rebel pickets by a portion of our party, while another succeeded in opening communication with Ringgold, some four or five miles distant.

For some of our friends at home, who do not understand the operations of war in detail, I will describe the business and duties of the Signal Corps, as briefly as possible. A code of signals, with flags in day time, and torches at night, has been established, by which messages are sent a distance, varying from one to twenty miles. This is done from the summit of high hills or mountain ranges by the aid of powerful glasses, when the distance is great. The motions of the flag to right or left, or front, are easily seen and read by the officer at the opposite station. Yesterday we cleared the brush from the top of a hill near by, and commenced "calling" the station at Ringgold. Calling consists in moving the flag back and forth until attention is attracted. This we done for two hours without success, until we hit upon the plan of making smoke; so we gathered a huge pile of pine boughs which soon sent up a smoke so black that in less than five minutes we had their attention and sent a message telling where the 4th Corps was, and asking where Gens. Thomas and Hooker were. Since then, messages have been sent quite often.

Our duties are light, and very pleasant, and we consider ourselves fortunate soldiers in being assigned to this branch of the service; still we are not anxious that the war should be prolonged on our account, for even to us "there is no place like home ...."

Yours, W.

— *Western Reserve Chronicle,* May 25, 1864

Eds. Chronicle: I am aware how anxiously those at home await news from absent ones in the army, and therefore I send you the following passages from recent letters received from Lieut. A.C. Dilley, Company C, 125th O.V.I. And yet what is it to receive news from a dear soldier, but the waking of a deeper desire to hear again; the opening wider of the lips of prayer for the imperiled one! We must all bear hearts schooled to hear the worst tidings, for to many of us how soon must such tidings come.

Mrs. H.A.B.[4]

Rocky-Faced Ridge, Ga.,
May 11th, 1864.

Dear Sister and Friends: We moved out from Cleveland, Tenn.,
the 3d inst., camped that night within 15 miles of Dalton, Ga.,
corps headquarters being at Catoosa Springs, 3 miles from Ringgold. Our corps (the 4th) moved to the right and camped on a
ridge, the name of which I do not know; our right connecting with
the 14th, our left with the 23d corps. Here we remained fortifying
till the 6th. On the night of the 6th, I went on picket. On the
morning of the 7th, our picket line was advanced until we drove in
the rebel pickets and pushed the rebels off Tunnel Hill, where we
camped that night.

At dawning of the 8th, we moved out — the 125th in advance.
After reconnoitering at the head of the valley, between the ridges
they occupy, it was decided to attack the steep ridge on the right.
Col. Harker, who never before said "can't," said that ridge could
not be ascended (so I hear), but Col. Opdycke said the 125th could
go it, and advanced us alone up the mountain side (which in
places was almost perpendicular) to a height of 600 feet. We
found them unprepared at that point, as they had not considered
that we all are borne on eagle wings in our patriotic mission, and
thinking inaccessible meant inaccessible to American soldiers as
to the rest of mankind, they were quite at ease. We captured one
sentinel who was so engaged watching our movements in the valley as not to have observed us passing upward.

We won the crest, which was only a few feet wide. Moving
along this a few hundred yards, we came on their works, Cos. E, F
and I being deployed, with C as reserve. Lieut. Phillips and myself being with I, made an effort to drive them from their works by
throwing ourselves on their left flank, where we suffered considerably, losing 2 killed, 5 wounded and one missing. Capt. Bates
brought C upon the crest, delivering a few volleys, which did not
half avenge their loss, Eli Swineheart[5] meeting his death, and six
other noble soldiers wounded.

We moved up within a few feet of their works, taking shelter
among rocks, where shots were given as either party showed their
heads. The rebels were behind a stone wall which was pierced for
their muskets.[6] They also sent stones over with their hands to
save their ammunition and dislodge us; and thus wore away the
afternoon of Sunday, we reading "sermons in stones," and sending

Second Lieutenant Alson C. Dilley was awaiting muster as first lieutenant of Company C when he was shot through the head June 27 at Kennesaw Mountain. An older brother, Lewis S. Dilley, served from August 1862 to June 1865 in the 103rd Ohio as an orderly sergeant, lieutenant and captain, and was an army correspondent for the *Cleveland Morning Leader.*

Clark, *Opdycke Tigers*

our prayers on bullet wings. In the evening we were relieved and fell back a short distance and rested that night. Our regiment on that occasion lost 24 killed and wounded.

That night two pieces of artillery were brought up by hand and timber cut away so that they could be used. In the morning we found their first line evacuated. Again we were moved up, deployed as support for the 27th Ill., and lay under fire until 5 P.M. Here Leonard Curtis was wounded, not seriously I think.[7] At 5 o'clock a part of our brigade charged the works, but mostly owing to General Wagner[8] not moving upon their flank, we were repulsed. The crest being wide enough for only 12 men abreast, and its sides nearly perpendicular, made a dreadful place to charge upon a full garrison protected by works. Co. I lost 2 killed, 6 wounded, and one missing; Co. C had 4 wounded. During both days our regiment lost in all about 50 men. Our brigade loss I do not know. Col. McElbane[9] [sic] of the 64th Ohio, killed; Col. Buckner,[10] 79th Ill., dangerously wounded; Lt. Col. Bullet,[11] [sic] 3rd Ky., mortally wounded; Maj. of 27th Ill., wounded; Lt. Col. Moore, 125th Ohio, in breast, slightly.[12]

On the 8th, while lying under fire on the crest, I saw the 23d corps advance on the enemy's works on the next hill to our left. They were checked by heavy artillery fire, and on the 9th fell back. While I write, the pickets a short distance in advance are skirmishing, and our artillery within five rods of me is shelling the foe at short range. We judge the enemy to be in strong force here, well fortified and intent on a fight; but our army is in good condition, has plenty of rations, and I think there is a movement going on at our right that will give us an advantage.

Fondly I am brother, son and friend,

A.C.D.

— *Western Reserve Chronicle,* June 1, 1864

Friend Converse:[13] Not through willful neglect have I thus far failed to fulfill my promise, made to you last winter, viz., that I would keep you posted as to the actions of the 125th during the present campaign. I now propose to redeem that promise, as far as may be, by giving you an historical account of those scenes through which we have passed, which will be of the most interest to readers.

The rebel General Johnston had collected a strong army in and around Dalton, for the purpose of resisting the progress of Sherman into Georgia. Having completed all the arrangements necessary to the moving of his immense force, and to the conducting of a long and active campaign, General Sherman gave that order, which we have so steadily and constantly obeyed ever since, the order of "Forward!"

The 125th had been, for some time, encamped near Cleveland, Tenn., but on the morning of the 3d of May, we struck our tents. Though the regiment had lately been largely recruited, yet through the unceasing and constant efforts of Col. Opdycke, Lt. Col. Moore and Major Bruff,[14] it was almost perfect in drill and gave bright promise of that success which it has since realized. We set out in the commencement of our history as a regiment with the motto our Colonel gave us, "Glorious Victory or an Honorable Grave," and our bright record and depleted ranks show that it is our motto still.

On the morning of the 8th of May, Colonel Harker was ordered by General Howard to send the best regiment in his brigade to dislodge a portion of the enemy from the position he had taken on a high and precipitous bluff called Rocky-Faced Ridge. The difficulties attendant upon the assaulting of this position were enough to appall the stoutest heart, yet Colonel Harker knew upon whom to call. The 125th was selected to storm and take the Ridge. Col. Opdycke, having formed the regiment by division, with a line of skirmishers in front, led on his men to the attack. Steep as the upright wall of a house was the face of the Ridge in places, while overhanging rocks and trees seemed to threaten with destruction all who should approach, or offered security to a concealed foe. We ascended from height to height along the only pass or road on the west side of the Ridge, which was so steep that our field officers were obliged to dismount and lead their horses. On we went, up, up, right in the face and eyes of the enemy. Nothing could daunt or check the upward march of the 125th, and we drove the enemy from their fancied security.

When we had gained possession of the northern part of the Ridge we were astonished at our success, for we were certain that we could have held it against any force of the enemy; and yet the 125th alone and unassisted had wrested a portion of it from a really superior adversary.

The top of the Ridge was not more than two rods in width, and in some places not over ten feet, with huge rocks piled up on either side, forming a most strong and impregnable natural fortification. After taking the north end of the Ridge, we drove the enemy along its summit, and our skirmishers advanced almost to their works on Buzzard's Roost. We threw up some hasty works across the summit of the Ridge, and bivouacked on the place. We lost in killed four, and wounded eighteen. So highly pleased was Gen. Howard that he issued a special order highly complimenting the 125th for their bold and daring exploit, and their success in overcoming and conquering the combined opposition of nature and the enemy.

During the following day, the 9th, the remainder of the Brigade came up, and it was determined to make an assault upon the enemy, who lay secure behind their works. Their fortifications were built so as to oppose any approach from whatever direction — and on the crest of a prolongation of the ridge we already held. At four P.M. the charge was made, the 125th being the second regiment, and in rear of the 64th O.V., who led the way. Part of our regiment was deployed as skirmishers, while the remainder was led to the attack, moving by the flank, for so narrow was the way along the ridge that four men could scarcely march abreast. Under a most murderous fire was the column pushed to within a few feet of the first line of works, when so destructive was the leaden torrent that the regiment in our front could stand it no longer, and came rushing, crowding back through the 125th, and actually forcing some of our men over the rocky ledge. In the effort to check the flying regiment, and to hold our own, Lt. Col. Moore was slightly wounded.

To succeed in pushing through the narrow gap, in the face of such a fire, and with a panic-stricken regiment forcing itself back through our ranks, was impossible. The whole Brigade fell back to the works which the 125th had built the night before. From deserters who came in that night we learned that Buzzard's Roost was held by two whole divisions of the enemy, while we assaulted it with but the Third Brigade. Had we taken the first line of their works, they could have fallen back about thirty feet to a second line; and, had we taken the second line, there was a third, and so on, line after line, and all filled with men. The ground was at such an inclination that, as we came through the pass, we received the fire of the entire force. No wonder, then, that one called it a "Leaden Sea of Death."

On the night of the 12th, the enemy evacuated his works, and, at an early hour, the army started in pursuit. We arrived at Dalton about noon. Pushing on after the retreating Johnston, we next came up with him on the 14th, and again found him behind strong works. His position was well chosen, occupying a range of hills and screened by heavy timber. The 125th, in advancing to the attack, had to pass over an open field exposed to a heavy fire of artillery. The field being passed, we had to wade through a deep stream and climb its steep banks, while bursting shell and crashing shot threatened to annihilate us. And, though the enemy threw shell right into our midst, yet the 125th faltered not, for Opdycke was at the head of his column.

One shell, striking in the regiment, wounded Col. Harker, and the command of the brigade devolved on Col. Opdycke, and left Lt. Col. Moore in command of the regiment. Pushing over the first line of works, Col. Opdycke wished to approach the second. Having jumped his horse over the works, he gave the command, "Forward!" and led the way to a ravine which, while it was much nearer to the enemy, afforded a good shelter from their fire. Ordering his men to lie down, he rode up on the hill or bank to survey more closely the actual position of the enemy. Scarcely had he exposed himself before he was the special object of attention of the enemy's sharpshooters, and he was almost instantly struck, the ball passing through his arm just above the elbow, and severely bruising his side. The wound was intensely painful, and he was compelled to give up and retire from the front.[15]

The right wing of the regiment, being nearer to the enemy than the left, was heavily engaged, and, under the command of Lt. Col. Moore, dealt death and wounds upon them. From 9 A.M. until almost dark did the gallant 125th rain a perfect storm of balls upon the foe, and successfully repulsed all attacks. Col. Moore was along the line time after time, and both he and Major Bruff, cheering and encouraging the men, moved as the controlling spirits of the storm.

We lost many in killed and wounded, and having exhausted our cartridges, we were relieved and taken back to replenish. This is the principal part we took in the fight near Resaca, for the enemy again took to their heels on the night of the 15th, and Sherman's army followed on the morning of the 16th in close pursuit. Thus ended the battle near Resaca, Ga.

But I have written enough for this time, and will close by promising that you shall hear soon again from

VICTOR, 125th Reg. O.V.I.

— *The Jeffersonian Democrat,* August 12, 1864

In Camp near Kingston, Ga.,
May 20th, 1864.

Dear Ones: The day is very hot and we are resting, as I assure you we have need to do. Our campaign has thus far been very severe, yet you will be glad to know, successful. Our advance has been a constant skirmish, our enemy taking advantage of every position to check us, which causes us to go forward in line of battle and take every precaution, as we well know we have an active foe, skillful in maneuver, watching every step, and making all they can from every ambush. Every stream has fortifications on its banks, from which our crossing is contested. Every hill they hold until we crowd them off. They post themselves within a border of woods, and rake us from safe concealment. This is their rear guard which seeks to gain time for their main army. Every evening we come upon them too late for a general engagement, and after some tedious maneuvering we get into camp. In the morning grey-coats are minus. We get some deserters and would get many more if they did not select their rear guard with such care, placing there the regular fire-eaters.

On the 12th we moved off the ridge, whence I wrote you on the 11th. At daylight we moved to the left, went on picket, then up to fortifications in front of the reserve. The following morning revealed to us that the enemy's works were unoccupied. We immediately moved in pursuit, halting half an hour in Dalton at 11 A.M., where everything was in confusion, just as our retiring friends had left things. Our boys foraged some here. One house full of pea nuts was soon "rifled" of its contents, and tobacco found there hushed many a hot tongue's craving.

Being two hours behind our foe, we pushed on again with vigor. We came upon their rear guard at dark about ten miles from Dalton, where, after some skirmishing, we camped. Saturday (the 14th) we moved two miles, where we found our *sang-froid flame-devourers* looking securely from their fortifications on the north bank of the [Oostanaula River] at Resaca.

The 23d corps moved in, our corps (4th) in reserve. About noon our brigade went in on the run, crossing an open field under [canister fire]. Advancing to the crest we relieved a part of the 23d corps, our regiment relieving the 103d Ohio, in which I saw many of our brother Lewis' friends.[16] On learning who I was, one from his company implored me for cartridges, seeming to think I could grant no favor so great as that. They were making a noble stand, but were out of ammunition.

Once more we found ourselves under a storm of fire, where we remained until our ammunition was used up. As the rebels occupied a side hill, with a ravine lying between us and their nest, we could not charge but were obliged to rely on bullets. Here in the fierce tumult of missiles a shell fragment struck Col. Harker in the leg, not wounding seriously, I believe. At this time Col. Opdycke was wounded with a musket ball in the arm, and though unable to do duty, insists on remaining with us. Here, company C had five wounded, and company I, 6. This campaign, so far, has cost us 97 men (our regiment, I mean), and the great battle is yet unfought, deferred only by their running.

Sunday was passed in severe fighting, mostly with artillery. Hooker charged their left flank, and captured their two outer lines of works. Our corps was not engaged. At midnight the reb-

Private Joseph H. Keys, Company F, was among 51 members of the regiment wounded at Resaca, six of whom died. A notation in the Bladensburg, Ohio farmer's service record reads: "Was engaged valiently [sic] in the battle of Chickamauga ... and in all the skirmishing of the company before and afterwards."

els made a general rush, but were repulsed. It may have been to silence suspicions of their departure; at any rate we found at dawn that they had evacuated as usual. They burned part of the railroad bridge here, and made an effort to do likewise with the wagon bridge, but our forces were so close as to save the latter. So little damage had been done the road before this, that our train was up before noon. This is what we call keeping "closed up." As I write, the cars are whistling, having overcome all obstacles. We are all in good spirits and hopeful.

I saw the 105th O.V.I. lately. All well, whom I knew. Mervin Tidd,[17] looking well; Clisby Ballard,[18] fat and hearty. Have not seen brother Lew, he not being with his regiment (103d), but is Post Commissary at Charleston, Tenn. Letters are not very plentiful here. Write! write! We shall receive if you only send. Duty calls. Good-bye.

Alson C. Dilley

— *Western Reserve Chronicle,* June 1, 1864

On the Field of Battle,
May 15th, 1864.

Eds. Chronicle: As many anxious hearts await the result of the battles now being fought in [Georgia], I take opportunity to send you a list of casualties in this regiment ....

In yesterday's fight we lost four killed, and forty wounded. Among the latter is Col. Opdycke, flesh wound in left arm ....

The battle is still raging furiously. While I write my regiment is under cover of a hill resting on arms, momentarily expecting to be marched again into the thickest of the fight. Yesterday we charged and took the first line of the enemy's works. It was, while leading a charge, waving his sword in advance of his regiment, that Col. Opdycke was wounded. General Harker was thrown from his horse and badly bruised by the bursting of a shell, just as we reached the enemy's works. Both Gen. Harker and Col. Opdycke are doing well; they rode along the lines to-day, encouraging the men; but neither will be able to resume command for several days. Gen. Hooker drove the enemy, capturing many prisoners and several cannon. Many brave men are falling; but the glorious old flag is being borne to victory ....

Respectfully,   CEYLON

— *Western Reserve Chronicle,* June 1, 1864

Resaca.

"... At 9 formed line (2d line, 65th O on my
right, 42 Ill on my left, 51 Ill to my front).
brig. soon moved to relieve Genl. Cox's
men. Severe fire — 125th went in
splendidly. Harker hit by a shell. I by
a minnie, all in left arm. Blead [*sic*]
profusely. fainted."

• Emerson Opdycke diary, May 14, 1864

In the Furnace,
Sunday morning 9:15 May 15

My Dearest Colonel — We are in bivouac, *rather,* lying beside
our arms in the woods to the left of which we debouched to charge
down the hill yesterday p.m., say 400 or 500 yards from the base
of that *terrible ridge.* The work goes on much as ever. There is an
incessant rattle of small arms with a heavy artillery bass. Our
men are busy as beavers throwing up a breast-work of logs & dirt.
We join on to 42 Ill. And almost every moment some one asks,
"How is Col. Opdycke this morning?" And when I answer them in
your words, especially these, "Tell the boys I would love to be with
them," their eyes glisten & they choke as they say, "I wish he could
be; he is the very man we would like to have with us — he & Col.
Harker, too." And my dear Colonel lies suffering this Sabbath day!
May the Lord of the Sabbath comfort him! O how thankful I am
that you escaped so well. Severe though your wound is, I shudder
to think how narrowly you escaped death; so that the escape from
actual death much reconciles me to your present suffering. Heav-
en, in whom you trust — Heaven grant you the full measure of
consolation.

My trust is in God. I hope & pray I may have grace and wis-
dom and fortitude for all the great responsibilities resting upon
me. You ought to go home till your wound heals.

D.H.M.
[David H. Moore]

— Ohio Historical Society

Near Dallas, Georgia,
June 4, 1864.

Eds. Chronicle: The rebels evacuated their works in front of
Resaca on the night of the 15th of May. The rear guard made a
charge at midnight in order to more effectually cover the retreat.
At daybreak next morning we discovered that we were confront-
ing empty intrenchments, and immediately commenced pursuit.
Details were sent to bury the dead and look after the wounded
who had fallen in the charge of the previous day.

I was curious to know how much we had injured the enemy,
and the extent of his works. Riding over the battle-field I crossed
a dense thicket where our men had charged a rebel battery. Such
a sight as met my eyes I hope never again to witness. After our
men had been repulsed, the woods had been set on fire, and the
dead and those too badly wounded to get away were literally
charred. The clothes of many had been almost entirely consumed
and their distorted countenances, and crisped and bloated forms,
expressed a hideousness altogether inhuman. Nearer to the rebel
works, friend and foe lay side by side mute in death, their blood
commingled. By observing closely, the remark I have several
times heard, seemed to be verified: "The rebel dead have a more
ghastly look than our own." While the countenances of many of
our men are composed, calm almost as wrapt in sleep, there is in
the faces of the enemy a troubled, ferocious, despairing look as
though death had increased rather than relieved his suffering. I
have never yet seen a placid countenance among the rebel dead.

Passing inside the enemy's works, I found he had been directed
to the "last ditch." We had driven him from his outer works, and
he had but one line left, nor would the character of the ground
permit another within supporting distance. This explained to me
why the 4th Corps had been so stubbornly opposed in all its attempts to charge the enemy's second line. Had we been successful
the result would have been disastrous to him. As it was we had
inflicted severe punishment. One battery had lost every horse but
one; and many new-made graves in the immediate vicinity attest-
ed how effective had been our fire. The fortifications were not
nearly so strong as those in front of Tunnel Hill; and had Johnston persisted in holding them one day longer, his army would
probably have been routed.

By 9 o'clock A.M. the 14th corps was in Resaca. In their haste to get away the enemy left valuable property behind. Flour, cannons, one thousand sacks of corn, tents, equipments &c., fell into our hands. All the intensely secesh citizens had left in advance of Johnston's army; and when our own advance entered the town, drums beating and banners flying, we were warmly welcomed. Small boys swung their hats and hurrahed for "Lincum," young ladies waved their handkerchiefs, and in the eyes of gray-headed men a tear-drop gathered as a sight of the honored emblem of our country awakened within their hearts old loves and old associations.

The railroad bridge was still burning; it had been fired some hours before. The braces of the wooden bridge had been nearly cut off, the enemy undoubtedly thinking that in our eagerness to pursue, our advance would rush madly on to the bridge when it would give way, and Yankees, bridge and all would be precipitated into the river. But our impetuosity was not so confused as to allow us to be caught in a trap of so shallow minded conception. After a little repairs we commenced crossing in single file, our brigade having the advance; and when our pioneers had made the bridge perfectly safe (the work of only a few moments) the entire corps passed rapidly over.

The advance was immediately resumed. At dark we went into bivouac near Calhoun having skirmished with the enemy's rear-guard all the afternoon, driving it steadily before us, capturing many prisoners and taking a large number of deserters and stragglers.

For the part taken by the 125th at the battle of Resaca, we claim no particular credit. We tried to do our duty; and at Rocky Face Ridge were highly complimented by the commanding General. Our loss was four killed and thirty-eight wounded. Among the latter was Col. Opdycke, whose wound I am pleased to inform you is now healing rapidly.

Respectfully, CEYLON

— *Western Reserve Chronicle,* June 22, 1864

Friend Converse: ... Pushing on after the enemy in his flight from his position near Resaca, our regiment advanced by the direct road to town, which we reached at an early hour. Here we found that the enemy had vainly attempted to retard our pursuit by burning the railroad and wagon bridges that span the Oostanaula river; and, though he succeeded in destroying the first, yet he failed with the other, for our line of skirmishers was so close upon him that they removed the fire before it had done much damage. The 3d brigade was first to cross the river, and we were continually marching and skirmishing with the enemy's rear guard until May 20th, when we were allowed a couple of days in which to rest.

On the 23d of May we again set forward, crossed the Etowah river, and continued our march with the greatest rapidity, bivouacking in the open fields at night, until the 25th. No words can express the severity of that march across the Allatoona hills, through heat and dust, now halting to form in line of battle as the enemy made a more stubborn resistance, and now pushing on again in column as that resistance was overcome. Halting when night came on, too weary to cook our supper, we would sink down upon the damp ground to rest and dream of home and home comforts.

Late in the evening of the 25th of May we moved in position to support the 20th Army Corps, which had found and assaulted the enemy, but suffered a repulse about one and a half miles from Dallas.[19]

For eleven days we were engaged in a constant and almost uninterrupted skirmish with the foe, whose position was too well chosen and too securely protected by heavy works to allow any hope of success by direct assault. But Sherman once more put his "Flanking Machine" in motion, compelling Johnston to leave his works or suffer an attack upon his flank.

Sabbath morning, June 5th, revealed the fact that our front was again clear, and we lost no time in following the enemy whom we again encountered at Pine Mountain. Here the usual cautious advance and the same flank movement obliged the enemy to retreat, which he did, falling back two miles into another line of works.

On the 18th of June Gen. Harker[20] again called for the 125th to take the advance, with the order to push as near the enemy's line as possible. Crossing a deep but narrow creek,[21] we advanced upon the enemy's skirmishers upon the run, driving them like sheep before us into their main line of works. Their artillery now opened upon us, but we continued to advance until within some forty rods of their works, when we halted and covered ourselves as best we could from the enemy's fire. There was a slight inequality, or rather unevenness, in the ground in our front, which saved us. Their artillery was soon silenced by our sharp-shooters, who were sent out to accomplish that object. Owing to our rapid yet cautious advance, we suffered but slight loss, having two killed and one wounded.[22]

The night of the 18th the enemy again fell back, leaving us his works, and on the morning of the 19th we started in pursuit. We well understood to what position he had fallen back, for right in our front was the frowning Kenesaw, whose base and summit glistened with bristling bayonets ....

VICTOR

— *The Jeffersonian Democrat,* October 21, 1864

Notes to Chapter Five

1. Rollin D. Barnes previously served as a corporal and sergeant in Company B. *Ohio Roster,* vol. VIII, p. 423.

2. Apparently the *Chronicle's* editor censored the number given by "W." According to General Sherman, on May 1, 1864 the Army of the Cumberland's effective strength was 60,773 officers and men. *OR,* vol. XXXVIII, pt. 1, p. 62.

3. After the war Sherman wrote that to make his forces as compact and mobile as possible, he issued strict orders "in relation to wagons and all species of incumbrances and impedimenta whatever. Each officer and soldier was required to carry on his horse or person food and clothing enough for five days. To each regiment was allowed but one wagon and one ambulance, and to the officers of each company one pack-horse or mule. Each division and brigade was provided a fair proportion of wagons for a supply-train, and these were limited in their loads to carry food, ammunition, and clothing. Michael Fellman, editor, *Memoirs of General W.T. Sherman* (New York: Penguin Books, 2000), p. 397.

4. H.A.B. was probably Henry A. Bell of Company C. He was promoted from private to sergeant major March 1, 1864 and wounded at Resaca. *Ohio Roster,* vol. VIII, p. 419.

5. Private Eli Swinehart, Company C, was shot in the head. A farmer from Defiance County, Ohio, he was 19 years old upon enlistment at Columbus December 10, 1863. His remains lie in the Chattanooga National Cemetery. Eli Swinehart CSR, RG 94, NARA.

6. The Confederates belonged to Brigadier General Edmund W. Pettus' Alabama brigade. Clark, p. 225.

7. Private Leonard H. Curtis, Company C, was discharged for disability in August 1865. *Ohio Roster,* vol. VIII, p. 427.

8. Brigadier General George D. Wagner commanded the 2nd Brigade, 2nd Division, 4th Corps.

9. Colonel Alexander McIlvain, 64th Ohio, was fatally wounded May 9 while attempting to assist an injured man. Wilbur F. Hinman, *The Story of the Sherman Brigade* (Alliance, Ohio: Press of Daily Review, 1897), p. 518-519.

10. Colonel Allen Buckner, 79th Illinois, was shot through the body but eventually recovered to resume command of his regiment. Hinman, p. 520.

11. Lieutenant Colonel William A. Bullitt, 3rd Kentucky, was shot through the thigh. He survived the wound but was disabled for life. Hinman, p. 520.

12. Moore wrote his wife May 10, explaining that he was "hit four times, once by a ball which passed through a corporal's head, struck me in the back of the hip and lodged in the lining of my blouse; that only *stung.* Another stripped my right coat sleeve below the elbow, a fragment of another hit me in the left breast; still another struck my right lower bowels. These two last hurt, but are only *slight;* they do not lay me up." The corporal mentioned, Luther S. Calvin of Company A, survived his injury. A bullet entered his left temple, passed downward to shatter his lower right

jaw, and exited before striking Moore. Clark, p. 230-231.

13. Julius O. Converse edited *The Jeffersonian Democrat,* published at Chardon, Ohio.

14. Joseph Bruff, captain of Company A, was promoted to major in February 1864 and mustered March 1. Joseph Bruff CSR.

15. Writing in third-person style after the war, Opdycke provided this account of his injury: "Col. Opdycke received a severe wound in the left arm just above the elbow, a jagged spent ball passing entirely through the arm, and lodging in the sleeve, having grazed the bone and nearly severed the principal tendon. Col. Opdycke fainted from loss of blood, and was taken from the field, but as soon as the arm was tied up to prevent a further loss of blood, he returned to his regiment, and remained with it until it was relieved .... The severity of Col. Opdycke's wound caused the Surgeons to earnestly recommend him to go into hospital, but he continued at the head of his command ...." "Military History of Emerson Opdycke," Emerson Opdycke Papers, Box 3, Folder 9, OHS.

16. At the time, Dilley's older brother, Lewis S. Dilley, was first lieutenant of Company E, 103rd Ohio, a regiment belonging to the 2nd Brigade, 3rd Division, 23rd Corps. *Ohio Roster,* vol. VII, p. 524.

17. Private Jeremiah M. Tidd, Company I, 105th Ohio.

18. Private Clisby Ballard, Company B, 105th Ohio.

19. On May 25 Hooker's three divisions attacked Hood's entrenched corps at New Hope Church, four miles northeast of Dallas, Georgia. Confederate musketry and massed artillery fire from 16 guns inflicted about 1,000 Federal casualties. Hooker's men immediately dubbed the place the "Hell Hole." Boatner, p. 219.

20. Harker received his commission as brigadier general June 7, to rank from September 20, 1863, the second day of battle at Chickamauga.

21. Mud Creek.

22. The regimental history lists casualties on June 18 as two killed and eight wounded, one of whom died. Clark, p. 268.

SIX

' If the 125th can go no further, there is no use in trying '

Continually forced to thwart Sherman's flanking maneuvers, General Johnston's Army of Tennessee had fallen back within 20 miles of Atlanta by the third week of June 1864. The Confederate commander established a strong defensive line west of Marietta along Kennesaw Mountain and connecting hills and ridges. Sherman's Federals probed for weaknesses, much of their effort expended in torrential rain. "We did not need to be told that the enemy was again entrenching," wrote a lieutenant in the 125th Ohio's Company B. "We could see them along the sides of this 'Gibraltar' of America .... Fully a week was spent getting into position and nearly all this time in wet clothes. We were fortunate if at night we could find a brier patch in which to make our beds. Officers and men alike were getting desperate to meet the enemy."

Sherman thought so, too. Considering Kennesaw to be "the key to the whole country," he decided to attack Johnston's fortified lines on June 27. The assaulting troops, including Newton's division to which the 125th belonged, were met "with determined courage and in great force." The regiment's attempt to punch through Confederate earthworks on Cheatham Hill, like those of comrades at points elsewhere, was stymied. The attack failed, with more than 2,500 Union casualties. The 125th Ohio lost nearly a quarter of its effective strength, as well as its brigade commander.

Sherman resumed his flanking movements after the bloody repulse, compelling the Confederates to withdraw across the Chattahoochee River into Atlanta's elaborate fortifications. Johnston was replaced as Army of Tennessee commander by General John Bell Hood, who immediately launched three heavy but fruitless and costly attacks against Federal forces at Peachtree Creek,

Bald Hill and Ezra Church. Sherman then laid siege to the city, shelling it through August while he leapfrogged parts of his command south to sever Atlanta's remaining rail supply lines and communications. Hood evacuated the city September 1, and the next day Union troops began their triumphant entry.

Camp near Kenesaw Mt.,
June 23d 1864.

For the past twenty days we have been pressing the enemy closely. Every day we drive him from his works, or gain some new advantage. But he is stubborn and our path is marked with graves and the blood of the wounded. Johnston is no sooner forced from one position than he falls coolly back to another equally strong, which he defends with equal tenacity. The condition of his works, which are uniformly strong, as well as the testimony of prisoners that fall into our hands, establish the fact that he has a large pioneer corps constantly at work constructing defences to his rear, which he may turn to good account in case of emergency. Johnston would be invincible to anything less than an army of veterans under the gallant Sherman. But his destiny is sealed. He may fret till his traitor heart breaks. Sherman has an engagement to fulfill in Atlanta, and all the hell commissioned hosts of rebeldom cannot swerve him from his purpose.

Our losses are considerable, but the punishment we are inflicting is even more severe. We stake our lives upon the issue, proud in the consciousness that it were better that all should die rather than the stain of treason be left to darken the fair character of the banner under which we have so long and so patiently fought. The army is in excellent health and spirits ....

CEYLON

— *Cleveland Herald,* July 6, 1864

Eds. Chronicle: The Army of the Cumberland is still a moving, working mass of men, full of energy, patriotism and fight. With Forrest, Wheeler and Morgan[1] in our rear, and Johnston in front, we manage to sustain a grave equilibrium, eating, drinking, sleeping and making merry at uncertain intervals, and in the interim paying our compliments to the foe as advantage or favorable circumstances offer. If we are many times successful, we are as of-

162

ten sorrowing over the graves of comrades slain; and while we thank God for victory, we grieve that it is the price of blood. Our joy in anticipation of a bright future — the reward for all their toil and suffering — is mingled with bitter tears in remembrance of the noble dead.

Among the losses lately sustained by our regiment I append the following as being of special interest to the readers of the *Chronicle:*

Lt. Freeman Collins, of Ravenna, killed by the bursting of a shell on the 19th inst., while the regiment was going into position near the base of Kenesaw Mountain.[2]

During a reconnaissance on the 23d inst., the 125th was ordered to support the skirmish line. Out of one hundred men engaged, Capt. Sterling Manchester, of Ashtabula, Nathan B. Hatch and Robert J. Rice, of Southington, were killed.[3]

We were informed to-day of the death of Jesse Sample, of Liberty, in hospital at Chattanooga, from a wound received at the battle of Rocky Face Ridge.[4]

Since the opening of this campaign the 125th has lost one hundred and forty-nine men in action, twenty-four of which have been killed. The loss of the entire army has not been proportionately great, but has been severe. The general health is good. We have plenty of rations and are well clothed. The determination to push forward to victory is in no way abated. With the memory of the dead to inspire the living, this army acknowledges no defeat short of utter destruction, and can be satisfied with no victory short of complete success.

Respectfully,   CEYLON

— *Western Reserve Chronicle,* July, 13, 1864

U.S. Field Hospital,
Four miles from Marietta, Ga.,
June 24, 1864.

Mrs. S. Manchester: It becomes my painful duty to inform you that your husband, Capt. Sterling Manchester, was on yesterday afternoon very dangerously wounded while engaged on the skirmish line. We were driving the rebels out of rifle pits, when he in urging on his men was struck by a minnie ball, which passed through his left arm just below the elbow, breaking one bone, and severing the main artery. The ball entered the left side, probably

Clark, *Opdycke Tigers*

Chaplain John W. Lewis

Clark, *Opdycke Tigers*

Captain Sterling Manchester

entering the left lung, but did not go in very deep. He walked back some distance, but bleeding freely became very weak and faint. After the doctor had dressed his wound I came with him to the hospital about a mile and remained with him all night. He was quite restless during the night and some flighty, but always answered us reasonably or asked for anything he wanted. He was much prostrated last night, but is much revived this morning. The doctor says his arm cannot be saved and it is very doubtful about his recovery. Just now the doctor comes in and says he must die. In this deep affliction I most sincerely sympathize with you, and trust you will be sustained.

He is now resting much more easily. He says: "Tell my wife that I did my duty to my Country, and die in a good cause." He insisted that we should tell him the worst and not flatter him, but though I told him the doctor said he could not recover, he seems to think he will.

The regiment will regret to lose him, and his Company seem to feel the stroke severely.

Yours truly,
J.W. Lewis[5]
Chaplain 125th O.V.I.

— *Ashtabula Weekly Telegraph*, July 16, 1864

[From the *Cleveland Herald,* July 25, 1864] — Many of our citizens were acquainted with the late Captain Manchester of the 125th, who fell on the 23d of June, near Marietta, Georgia. Captain M., just in the morning of life, had just entered upon a flattering professional career when the war broke out. He entered the service as a private, was promoted rapidly and having served out two periods of service, he entered the third time, as Captain in the 125th. When in Ohio about six months since, he married a Miss Alvord of Ashtabula, and returned to his command with a heart only the more intensified with love of his flag, and with a determination only the more strengthened to put down this rebellion.

The news of the death of Captain M. reached his friends by dispatch. Two or three letters written by the deceased reached his wife after the wires had given her the sad news of his death, and after his remains had arrived at Ashtabula, the young widow received an unfinished letter saturated with her husband's blood, which was found in his pocket after receiving his death wound.

The funeral of the brave Captain was attended at Ashtabula on Saturday last by a large concourse, and every demonstration of respect was shown the gallant dead. The last words uttered by the noble Captain were: "Tell my wife that I did my duty to my country, and died in a good cause."

Lieut. Col. Moore has written the following letter in relation to the death of Capt. Manchester:

Bivouac near Marietta, Ga.,
June 24, 1864.

My Dear Mrs Manchester: This letter, addressed to you in a strange hand,[6] will doubtless fill your heart with unutterable anguish. God hath seen fit to call your beloved husband home. He who doeth all things well hath taken Capt. Manchester from earthly trials and struggles to eternal rest.

I look upon that war-worn glass, and haversack, and sword, stained with the life-blood of the gallant departed, and my very soul is whelmed with sorrow. It seems impossible to realize it; he who was but yesternoon with us in all the buoyancy and strength of vigorous manhood, came back to us yesternight pale and bleeding, and went away from us borne on a litter, never to gladden us again by his coming.

The skirmish line came drifting back, and still back; and the rebels came rushing on, nearer and still nearer. To me was confided the support of that line. Captains Moses, Stinger [7] and Manchester were sent out with their companies. Never was nobler behavior, never more conspicuous courage. Away in the far front I saw the captain cheering and urging on his men. I saw his men emulating his courage and zeal; and then the storm grew deeper, and the smoke shut the captain's form from view. It is not long till I see him again. Emerging from the battle clouds he comes with pale face and unsteady, faltering steps towards me. My God! he is wounded! With the help of another, I lay him tenderly upon a stretcher; remove his sword, haversack and fieldglass, the weight of which burden him; unloosen his coat and vest, moisten his lips, and, committing him to God, send him to the surgeon.

The tidings fall heavily upon my ear — *"Capt. Manchester is dead!"* That Capt. Manchester is dead is whispered from lip to lip, and the regiment grows sad and still as one tells another of the soldierly excellencies of the departed. But he died a hero. He died for his country ....[8]

Your sorrowing brother,

D.H. Moore

— *Cleveland Herald,* July 25, 1864

Friend Converse: Passing minor incidents that transpired, we come to the history of the 27th of June, a day never to be forgotten by those who participated in its terrors before Kenesaw — one of the most bloody days in the history of the 125th.

I do not presume to be able to give you anything like a description of the terrible grandeur of the scene that occurred around Kenesaw Mountain, Georgia, on the eventful 27th of June, for it is beyond my power to paint in words the sweeping line, the rushing column, and the deadly charge. The army of Johnston had been forced back, step by step, ever retreating but fighting still, until their lines reached one of the strongest positions, both naturally and artificially, which it is possible to imagine. Their right rested on rocky and precipitous Kenesaw, their left crossing the town of Marietta, while their whole front was protected by massive and almost impregnable works.

For reasons best known to our generals, it had been determined that a general assault should be made upon portions of the

Mass. MOLLUS, USAMHI

Federal trenches filled the foreground of a view showing Little Kennesaw Mountain (left) and Pigeon Hill. Newton's 4th Corps' division assaulted Confederate breast-works south of this position.

enemy's works, and the morning of the 27th selected as the time of trial. At an early hour the 125th was moved out and taken to that part of our line occupied by Stanley's forces,[9] and was here deployed as skirmishers, crossing the front of our (the 3d) Brigade, and joined on the right and left by other regiments from other Divisions, the whole skirmish line being under the direction of Colonel Opdycke. The command of the 125th devolved on Lieut. Colonel D.H. Moore, aided by Major Bruff. Our line was formed immediately in rear of our outer works, and here waited the signal that was to send us forward like a living, crushing avalanche upon the foe.

As we lay deployed, I took a hasty view of the ground over which we were to pass. Immediately in front of our left, where I was stationed, was an open field extending down to a piece of timber, some forty rods distant, in the edge of which were the enemy's rifle pits, which were filled with the best troops of Johnston's army.[10] About fifteen rods in rear of those pits was the main line of the rebel works, and within range of our own. The right companies were similarly situated, while the centre of the regiment was more protected by a belt of timber which extended nearly to the enemy's line. It was not over fifty-five rods[11] from our breastworks to those of the enemy, and in passing over this space we were not only exposed to a fire from the rifle pits, but also that of the main works. Such was the nature of the ground that there was no shelter from the instant we crossed our works to go forward until we reached these rebel pits. Our orders were to "take those pits." Once over our own works there could be no safety but in instantaneous movement and success.

Col. Opdycke commanded "Forward" and the 125th, led by its gallant Lieut. Colonel and Major, had passed the last works that could shelter them from the enemy's fire and rushed upon the foe. The enemy's fire was most terrific. Secure behind their works, their infantry deluged us with balls while their artillery swept our depleted ranks with grape and shell, pouring upon us an unceasing tide of death. Many of our brave boys fell close to the enemy's abatis.[12] Lieut. Dilley fell dead within a few feet of the works while gallantly leading on his men.[13] Lieut. Burnham fell pierced by four balls close to the ditch.[14] Capt. Moses was badly wounded at the head of his company.[15] No line of skirmishers could storm those works in the face of such a fire. The assailing columns advanced to support us, but could not reach us, so severe was the fire.

At this instant General Harker fell from his horse mortally wounded;[16] the regiments on the right and left of the 125th fell back, drawing upon us the whole fire of the enemy in our front. No troops could live under such a fire, and the order came for the 125th to fall back to the captured pits and to hold them. Slowly and sullenly we fell back, and did hold these pits, though the enemy came out and made the most desperate attempts to wrest them from us. But in vain, for every pit was a "Tiger's lair," held by men who would have perished rather than have yielded them.

We had accomplished all we were ordered to do in taking and holding the rifle pits, for it was not expected that a skirmish line could successfully storm such main works as those of the enemy. The storming columns had been withdrawn before the 125th was ordered back to the rifle pits, and the attempt abandoned for the time. In due season the 125th was released and what remained of it was taken back to the ground occupied by it before starting out in the morning. Our loss was severe, numbering forty-four

Brigadier General Charles G. Harker, the 125th Ohio's 28-year-old brigade commander, died June 27 five hours after being unhorsed by a bullet that shattered his left arm and penetrated his side. Many in the regiment lamented his death, but none more than Opdycke, who called Harker "brave and purehearted. He was a noble sacrifice, offered for the noblest of causes."

Richard W. Nee Collection

men and nine officers killed and wounded. The almost unprecedented loss of officers arose from the fact that we were so close to the enemy that it was an easy matter for them to single out officers from the men.

A few instances of personal daring and address: As we were rushing upon the rifle pits, a Rebel officer, seeing Lieut. Colonel Moore in advance of the line, sprang toward him and, with drawn saber, demanded his surrender. But said rebel had miscalculated his man, for in an instant Col. Moore's revolver was at his breast, coupled with a demand for him, the rebel, to surrender. Col. Moore's argument being the more powerful, the rebel gave in and yielded his sword to the Colonel, who sent him to the rear.

In the heat of the engagement I had to pass along the line to execute an order which had been given me, and in so doing found Colonel Opdycke mounted on his favorite horse, riding as coolly along the skirmish line as though on a pleasure tour, instead of being in the midst of a storm of balls. I also saw a private of Co. B, Isaac Brown, rush into a rifle pit and capture its entire contents of five rebels. These instances came under my own observation. No doubt but there were many other deeds of personal daring, for it was a day of daring deeds.

A truce was shortly after agreed upon for the purpose of allowing us to bury our dead, which we did, and they now fill soldiers' graves. "Brave boys were they," but their fighting is now over and their names are added to the list of those who fell in their country's cause. The motto of the 125th is "A glorious victory or an honorable grave," and it is a common saying here in the 4th Corps that where "Opdycke's Tigers" cannot go, no other troops need try.

VICTOR

— *The Jeffersonian Democrat,* October 21, 1864

P.S. — Col. Opdycke, though still suffering from a severe wound received at Resaca, has never left the brigade. A little anecdote here comes in. The day after he was wounded, he was earnestly requested and advised by the surgeon to go to the rear, as fears were entertained that he would lose his arm. The Colonel at last consented, and went back four miles, but in a few hours he came back to the regiment, saying, "Boys, I am ashamed to leave you with such a little hurt."

VIC.

— *Western Reserve Chronicle,* July 27, 1864

In Bivouac South of the Chattahoochee,
July 28th, 1864.

Eds. Chronicle: There are many details connected with the charge of the 27th of June that will lose none of their interest on account of intervening time. In this brief letter, it is only intended to give such facts as it is hoped will be especially welcome to the friends of the 125th in good Old Trumbull.

Of the extent and object of the charge, you are already informed. As far as accomplishing the desired end is concerned, it was a failure. But it was a glorious failure, developing a bravery unsurpassed and unsurpassable. Men never displayed a more heroic courage; nor did they ever yield with more reluctance an uneven contest — a contest of courage against science, where men

entirely unprotected were thrown against heavy works guarded
by every device known by modern warfare, where no foe could be
seen but where the very air was filled with death-shots from un-
seen hands.

In ordering the charge, General Sherman doubtless thought
the enemy had materially weakened his lines. Had he estimated
correctly the result of the charge would have been the overthrow
of Johnston's army, and the nation's heart would have been
thrilled with joy. As it is, a few scores of graves, a few hundred
brave men maimed for life have been the occasion of increasing
our sorrow, but have in no way diminished our confidence of final
victory.

The plan was comprehensive. Four veteran divisions were
chosen for the onset. Newton's (2d Div.) of the 4th corps, composed of three brigades, formed three columns of attack, thirty
lines deep, each regiment being in column by division. Harker's
(the 3d) Brigade on the right was drawn up before an angle of the
enemy's works where he had massed heavily both artillery and
infantry.[17] Colonel Opdycke was in charge of three regiments (ordered to be the best in the division) to be deployed as skirmishers
in front of the three assaulting columns. The 125th was selected
from the 3d brigade. We had to cross a ravine and ascend a gentle
slope covered with a thick growth of small trees three hundred
paces before reaching the main rebel works. So close were we to
the rebel rifle pits that it was necessary to give the preparatory
commands in an undertone to prevent them from being over-
heard.

Arrangements were rapidly consummated, and at the appointed time we were ready. A short prayer imploring victory ascended
to the God of battles, a slight premonition of danger was brushed
from our minds, and we waited the signal. At the sound of the
bugle we bounded over the breastworks and charged with a yell.
The next moment we had captured the entire rebel picket-line in
our front (some thirty men in number, including three officers)
and were again advancing. We had scarcely cleared his rifle pits,
however, when we received a withering volley from the enemy's
main works, staggering us for a moment, but in no way discom-
posing our line. Several men were badly wounded, but they were
left to the care of the stretcher-bearers and the line pressed forward even more rapidly than before.

Arriving within five rods of the rebel works, we encountered a

Sergeant William G. Weimer, Company I, was fatally shot during the June 27 Kennesaw Mountain assault. The 24-year-old Muskingum County farmer was one of 17 officers and men from the 125th killed or mortally wounded that day. His remains lie in Marietta National Cemetery.

heavy abatis which it was madness for a skirmish line to attempt to pass. The men threw themselves flat on the ground and commenced to load and fire, always rising to take aim, as fast as possible. The head of the storming column had no sooner come up than it, too, dropped to the ground and commenced firing by detail. Soon the entire column had come to a halt closing in mass within six rods of the enemy's works. Some of the men were heard to say, *"If the 125th can go no further, there is no use in trying."*

Our fire now became more rapid than that of the enemy. He dare not raise a head above his works, but fired at random, being guided by his knowledge of the ground and our near approach. Thus we stood full twenty minutes facing the muzzles of rebel guns, exposed to a terrible fire, but unable to proceed. Could a simultaneous charge have been made we might, perhaps, have carried the works before us. The enemy were already half cowed. No command could be heard beyond a few feet. The roar of musketry was incessant. A cloud of smoke rising from the discharged guns obscured the sun and enveloped us in darkness — *the terrible night of battle,* such as we had seen before, but such only as soldiers can comprehend.

At this juncture General Harker rode along the lines shouting to the men to charge. But it was in vain. The ground was already strewn with the dead and dying, and the men were fast losing heart. Many were willing enough to attempt to proceed, but each seemed to doubt the other, and no one would advance more than a few feet without again dropping on the ground. General Harker now rode through the ranks swinging his hat and urging the men by every inducement to push forward. A score or more of brave hearts accustomed to obey their loved commander at every hazard sprang to their feet. The colors of the 27th Ill. were planted on the rebel parapet. The 3d Ky. color Sergeant dashed boldly forward, and several others struggled through the dense abatis. But it all only served more fully to show how fruitless must be every attempt. General Harker fell from his horse and was borne from the field mortally wounded.

The 27th Illinois color bearer was pierced by half a dozen balls and his colors fell inside the rebel works.[18] The 3d Ky. color bearer was mortally wounded and his flag was taken up by Benjamin Porter [19] of the 125th, and carried through the rest of the fight. Every man who attempted to scale the rebel works was either

Inventory of the effects of Alson C Dilley late a 2nd
Lieut of Capt Edward P. Bates' Company "C" 125th Regt
Ohio Vols who was enrolled as a 2nd Lieut at Columbus,
Ohio on the 3d day of March 1864, and mustered into the
service of the United States as a 2nd Lieut on the 5th day
of March 1864 at Cleveland Tenn in Co. "C" 125th O.V.I.
to serve three years or during the war. He was born in
Howland Trumbull County State of Ohio. he was 24 years
of age, 5 feet 6 inches high, Fair complexion, blue eyes, black
hair, and by occupation when enrolled a Farmer. He died
[in] battle near Kennesaw Mountain Ga on the 27th day
of June 1864 by reason of Gun Shot wound in the head.

Inventory

Articles		Articles		Articles	
Uniform Coats	1	Port Folios	1	Wool Blankets	1
Vests	1	Pocket Books	1	Watch & Chain	1
Trowsers Pairs	1	Memorandum Book	1	Pin Cushion	1
Caps	1	Bibles	1	Compass	1
Swords	1	Casey's Tactics	1	Haversack	1
Sword Belts	1	Photographs	1	Valise	1
Sashes	1	Mirrors	1		
Gloves	1	Lead Pencils	3		
Handkerchiefs	3	Tooth Brushes	2		
Shirts	2	Combs	2	Bills	$3.15
Stockings	2	Pocket Knives	1	Notes	$45

I certify that the above is a correct Inventory of the effects
of Lieut. A.C. Dilley and that they are in the possession
of his legal representatives.

Atlanta Ga   Edw P. Bates
Sept 13th 1864   Capt 125th O.V.I.

killed or wounded. Our own brave Lieut. Dilley fell in advance of
his company within three rods of the enemy's works, shot through
the head. I can find no more appropriate place than this to speak
of his noble qualities. Brave to a fault, he seemed to fear no dan-

First Lieutenant Ephraim P. Evans, Company D, was shot in the right thigh at the outset of the regiment's June 27 charge. Married for 10 years, the former Ravenna carriage maker informed his wife, Eliza, by letter July 5 that the bullet in his leg could not be extracted until swelling subsided. He expressed hope of being home by August 1, but she never heard from him again. Evans died July 8 and was buried in Chattanooga National Cemetery.

Clark, *Opdycke Tigers*

ger; generous and kind on all occasions and under all circumstances, none knew but to love him. Though he may not stand so high on the catalogue of fame as many who have fallen in this most unholy war, yet among the pure spirits of a nobler clime he will be welcomed to a seat near to the eternal throne.

The enemy beginning to obtain an enfilading fire from our right, we were now forced slowly to fall back. We withdrew, however, without confusion, halting when we had reached the ravine and defying the enemy to leave his works and meet us on open ground. Not a rebel dare venture to follow. The 125th being the first to charge the rebel works was the last to give them up. We occupied his rifle pits half an hour after the column had withdrawn and until relieved by fresh troops. We carried with us most of our dead and wounded; a few who fell on the very slope of the enemy's works we were unable to get. In this condition was Lieut. Burnham of Kinsman, wounded in the thigh. After he was so badly hurt as to be almost helpless, he was made a target for rebel sharp-shooters and was shot no less than four times before he could drag himself out of range.

Lieut. Evans of Ravenna, severely wounded in the thigh, has since died of his wound.[20] At last accounts Capt. Elmer Moses and

Lieut. Burnham were doing well but their recovery is yet quite uncertain. Both of the above named are noble officers, and we miss them greatly in the regiment.

Capt. Moses (at the time a sergeant in the 41st O.V.) was wounded at the battle of Shiloh. His experience and courage have been of great benefit to the 125th. We feel under obligation to him and deeply regret that he will probably never again be able to be with us. Cassius Coates, of Mecca, has suffered amputation of his left arm, and is getting on finely.[21] P. Welch and Avery Harwood of Farmington, Howard Bascom of Greene, George and Sylvester Waterman of Mecca, and A. Hager of Brookfield are all doing well.[22] We are pleased to hear that all our wounded are now receiving good treatment at the rear.

We expect to make another advance to-morrow. Our ranks are greatly depleted. The excessive hot weather is telling upon those who are left. We do not shoulder as many muskets, nor is the step of our brave boys as elastic as formerly. But our courage is no way impaired. Love of country burns as bright in every heart, and our cause is none the less sacred because it is costing us dearly.

Respectfully,
CEYLON

— *Western Reserve Chronicle,* August 3, 1864

Friend Converse: After [the Kennesaw] assault we lay behind our works until July 3d, when the enemy evacuated his works, giving up to our possession the town of Marietta, and, throwing himself across the Chattahoochee river, made such a disposition of his forces as to dispute our following.

While the right of our army engaged the attention of the enemy by false demonstrations and skillful manoeuvres, thus leading him to look for an attempt to cross at that point, the left swung around and did cross with but little opposition.

July 18th found the 125th again in the front on the skirmish line. The rebels were more obstinate this day than usual, being encouraged, no doubt, by the presence of some light artillery on their skirmish line, and seemed determined not to yield.[23] The brigade skirmish line, being under the immediate direction of Colonel Opdycke, was pushed steadily forward and the enemy driven from his position. At Nancy's creek, where the rebels had thrown up some hasty works, an unusual and most determined

Brigadier General John Newton led the 4th Corps' 2nd Division throughout the Atlanta campaign. He previously served two years in the Army of the Potomac, and temporarily commanded the 1st Corps at Gettysburg. "Quick, firm and correct" were words Opdycke used to describe Newton upon their first meeting in April 1864.

Mass. MOLLUS, USAMHI

resistance was offered to our advance, when Col. Opdycke, dashing to the front, ordered the bugles to sound the charge. He had scarcely given the order when a volley of musketry was fired at him by the rebels, and his horse, staggering a few paces, fell dead.[24] Almost miraculously escaping death, he mounted a fresh horse and, our line moving up, the enemy was forced from the field. By our rapid and impetuous advance we drove the enemy about seven miles that day, and, halting as night closed upon us, we fortified our position.

July 20th is celebrated as the day on which the new rebel general, Hood, attempted for the first time to measure arms with Sherman, and was most signally defeated. Hood had allowed, as he supposed, but a small force to cross Peach Tree Creek, a deep and miry stream, and, being confident that he could easily capture and destroy this small force, he left his works and attacked us with the greatest impetuosity. Contrary, however, to his calculations, the greater part of our force had crossed, and he came squarely and fairly upon the 4th and 20th corps. It was a fair, open field fight, and was most bravely contested by both armies. Charge after charge was given and received, and for a few hours victory seemed to hang doubtful. But "Sherman's Flankers" and his "Abolition Sharp-Shooters," as the rebels call the 20th and 4th corps, proved too much for Southern chivalry and they retreated to their works, leaving us in possession of the field. The ground was covered with dead and wounded rebels and Union soldiers. In numerous instances I saw them lying side by side — proving the severity of the conflict.[25] The loss of the 125th was not severe.

Driving the enemy into the outer defences of Atlanta, our lines began to close around the fated city on the 22d of July, and on the morning of the 23d the siege had opened. The point which the 125th occupied was distant from the city about three miles. Heavy works and defences were thrown up by our army to protect us from assault, and, our heavy siege guns having arrived, the town was for a time severely bombarded ....

VICTOR

— *The Jeffersonian Democrat,* October 21, 1864

Camp near Atlanta, Ga.,
Sept. 20, '64.

Friend Converse: ... From July 22d to August 25th, our regiment was engaged in assisting to conduct the siege, and in constructing a large fort, to which was given the name Fort Opdycke, in honor of Colonel Opdycke, under whose guidance and superintendence it was built.[26]

On the morning of August 25th we were ordered to prepare for a march, and at 11 P.M. moved out of our works, thus commencing that grand flanking movement which resulted in the overthrow and capture of the Gate City of the Confederacy. Passing around to the right of the army, we continued our march until 6 A.M. when a short rest was given to allow the men an opportunity to get breakfast. At 8 A.M. August 26th, we again moved forward, continuing the march in the most rapid manner until 4 P.M., when we bivouacked for the night.

This day's march was one of the most severe in the whole campaign. It was necessary for the army to move with the greatest dispatch of which it was capable in order to reach the Macon railroad, the great goal of the expedition, before the enemy should discover our intention. Never before had I seen troops so completely exhausted as on the 26th. The heat was most intense, and numbers in the 125th fell under the effects of sunstroke, while others, worn out by the continued rapid march, were compelled to halt and lie down by the roadside to rest. On our making a short halt about 1 P.M. the entire regiment could boast 28 men present to stack arms, and three line officers. During the evening, however, after we had bivouacked, the absent came up again.

August 27th, 28th, 29th, 30th and 31st were occupied in continual marching and efforts to wreck the railroad, which we succeeded in doing on the 1st of Sept., striking it three miles below Rough and Ready, about 21 miles south from Atlanta. I can assure you we struck it a hard blow, tearing up some 12 miles of track, burning the ties and bending the rails in such a manner as to render them entirely useless. We were proceeding with our work of destruction when, about 4 P.M., we suddenly received orders to hasten towards Jonesboro, in support of the Army of the Tennessee, which had attacked and gained advantage over the enemy. We hastened on and, when the shades of night were fast closing around, we formed in position and charged upon the enemy's line, driving them after a short but severe struggle from the field. Our lines advanced to the edge of a wood formerly held by the enemy, and, darkness having come on, and, not knowing what might be in our front, we halted, sending out pickets, and rested for the night.

In establishing our picket line, Capt. R.C. Powers of the 125th discovered in our immediate front a rebel hospital, and, throwing forward the chain of pickets, he enclosed the said hospital and all its inmates within our line. Taking with him a guard of four men, he approached the hospital for the purpose of reconnoitering, but left his guard a few paces distant. Going up to the door of the tent, he saw it filled with wounded and sick, and, seeing also an attendant, called him. The rebel nurse advanced toward Capt. P., mistaking him in the darkness for one of his own officers. As they yet knew nothing of being within the Yankee lines until close up to him, when discovering the blue uniform, he began a backward movement, crying out at every step, "Oh, don't! Oh, don't!" Captain P., after a few words, convinced him that he did not intend to, and that he simply wished to ask him some questions.

While engaged in so doing, the surgeon in charge of the hospital approached — a little, stiff, pugnacious fellow, and *graciously*

Two of Emerson Opdycke's closest wartime friends were fellow residents of Warren, Ohio. Lieutenant Colonel Henry G. Stratton (top) and Captain Oscar O. Miller belonged to the 19th Ohio Infantry, a regiment assigned to the 4th Corps' 3rd Division during the summer of 1864. Both officers shared Opdycke's business and political beliefs, often dined with him in permanent camps, and frequently were mentioned in the colonel's correspondence to his wife. As a brigade adjutant, Miller was killed September 2, 1864 on the skirmish line near Lovejoy's Station, Georgia. "I cannot tell you how deeply I feel the loss of this brave and gallant officer," Opdycke sadly wrote home a day later.

informed Captain P. that he was mistaken, that he was within the rebel lines; and attempting to seize his (Capt. P.'s) sword, desired him to surrender. The other attendants had by this time assembled, armed in such a way as to encircle him. But, quickly stepping back a pace or two, Capt. Powers disengaged his sword and called on his guard to advance. The sight of Yankee bayonets, and the disclosure of the existing facts, soon converted the little fighting surgeon into a very humble, submissive prisoner, and he was speedily transferred to more secure quarters.

During the night the rebel army again fled; and, on the morning of the 2d, we passed through the town of Jonesboro in pursuit, and came up with him at Lovejoy Station. The enemy declining an open field fight, and being protected by strong works, and the commanding General having received positive and reliable information that Atlanta was occupied by our troops under General Slocum,[27] we commenced our march back to that place on the 5th. Moving at a very moderate rate, we reached the city on the 8th and went into camp two miles north-east of it on the Augusta railroad.

Thus gloriously has this long campaign, extending over a period of four months, ended; and the 125th has the proud consciousness of knowing that it has done its part. It has, on several occasions, been complimented by the Generals commanding the Brigade and Division for its prowess and daring. During the campaign we have lost heavily in killed and wounded, the records showing that we have had 39 killed and 213 wounded. Col. Opdycke is now in command of the 1st Brigade, 2d Division, 4th Army Corps, and the 125th has been transferred from the 3d Brigade to his command ....[28]

Your most ob't. &c.
VICTOR, 125th O.V.I.

— *The Jeffersonian Democrat,* November 4, 1864

Notes to Chapter Six

1. Confederate cavalry commanders Nathan Bedford Forrest, Joseph Wheeler and John Hunt Morgan.

2. Second Lieutenant Freeman Collins, Company D, was killed instantly only a few yards away from Opdycke, who wrote that "We were under a severe artillery fire and a peice [sic] of shell passed through his body, he was a good patriotic man." Collins had been an officer for little more than three months. *To Battle for God and the Right,* p. 185; *Ohio Roster,* vol. VIII, p. 429.

3. Manchester was captain of Company K. Privates Hatch and Rice, Company B, were from Trumbull County.

4. Private Jesse Sample, Company C, died May 29, 1864. He is buried in Chattanooga National Cemetery. *Ohio Roster,* vol. VIII, p. 428, 773.

5. The Reverend John W. Lewis was appointed the 125th Ohio's chaplain in November 1863, but did not join the regiment for duty until February 24, 1864. On January 13, 1865 he tendered his resignation, "Being unwilling to remain in a position where I find myself unable to render that service which I think [is] required by my Office." In forwarding the request up corps chain of command, Major Bruff of the 125th endorsed it "Approved. This officer is entirely worthless to the Regiment. He has neither the respect or confidence of officers or men, and his influence is bad." Undaunted, Lewis became chaplain of the 45th Ohio a month later and mustered out with that regiment in June 1865. John W. Lewis CSR, RG 94, NARA; *Ohio Roster,* vol. IV, p. 321.

6. Lieutenant Colonel Moore had attended the Manchesters' wedding in late 1863, shortly before Moore and Company K left Ohio to join the regiment. *To Battle for God and the Right,* p. 188.

7. Captains Elmer Moses and Daniel A. Stinger commanded Companies B and G, respectively.

8. Manchester's body was shipped by rail to Ashtabula, Ohio, where it arrived July 9 and lay in state in the mayor's office at town hall. A church funeral was held the next day, "the house packed to its utmost capacity." During the graveside procession "a shower came up and the rain began to fall so freely as to drive many to fall out, and seek shelter from its drenchings. As a consequence the Masonic ceremonies were shortened, and the assemblage compelled to withdraw." *Ashtabula Weekly Telegraph,* July 16, 1864.

9. At the time, Major General David S. Stanley commanded the 1st Division, 4th Corps.

10. The point of attack for Harker's brigade was near the junction of Confederate divisions commanded by Major Generals Benjamin F. Cheatham and Patrick R. Cleburne.

11. A little more than 300 yards.

12. Abatis was a defensive barrier of fallen trees with branches, sometimes sharpened, pointed toward the enemy.

13. Dilley was shot through the head. Alson C. Dilley CSR.

14. Sergeant Thomas M. Burnham, Company B, received gunshot

wounds in the right shoulder, right hand and right hip. He was promoted to second lieutenant in May 1864, but still was awaiting muster when he died at Chattanooga July 13. His body was returned to Trumbull County, where he farmed prior to his enlistment. Thomas M. Burnham CSR, RG 94, NARA.

15. Moses was disabled for the rest of the war when his left thigh was fractured. A full year after the Kennesaw assault he informed Opdycke that "I am improving very slow. My wounds are not healed yet and now and then I get a piece of bone. I have seven pieces in all, and I am satisfied there is more to come. I can bear considerable weight on my limb but not enough to go without crutches; my knee is stiff and the Surgeon tells me it always will be and my limb is full 1 and 1/2 inches shorter than the other." Moses to Opdycke, June 25, 1865, Emerson Opdycke Papers, OHS.

16. According to Opdycke, Harker fell "trying to inspire his men. The fatal ball had shattered his left arm, and entered [his] side .... I saw him and talked a few hurried words with him. He expired easily at 1 P.M." *To Battle for God and the Right*, p. 189.

17. This salient in the Confederate line became known as the "Dead Angle," and was occupied by Tennessee troops of B.F. Cheatham's division. Harker's brigade attacked just north of the angle.

18. Color Sergeant Michael Delaney, Company K, 27th Illinois, planted his regimental flag on the earthworks' parapet when he was shot in the face and bayoneted in the upper chest. The flag and staff fell into the works as Delaney staggered backward. He died July 9. *Illinois AGR,* vol. II, p. 411, 416; Hinman, p. 548.

19. Private Porter belonged to Company I.

20. First Lieutenant Ephraim P. Evans, Company D, was born in Wheeling, Virginia, and moved to Ravenna, Ohio, in 1853, where he married the next year and resided until his enlistment. Evans died July 8 at Chattanooga. Among his effects were a gum coat, two gold pens, two pocket knives, a deerskin robe and $7.25. Ephraim P. Evans CSR, RG 94, NARA.

21. Private Coates, a February 1864 enlistee in Company C, was discharged for disability in February 1865. *Ohio Roster,* vol. VIII, p. 426.

22. All the privates mentioned belonged to Companies B and C. Asa Hager and Sylvester Waterman died of their wounds July 13 and 18, respectively. *Ohio Roster,* vol. VIII, p. 424, 428.

23. Lieutenant Colonel Moore reported the artillery was a four-gun battery of howitzers supporting dismounted cavalrymen of Wheeler's command. *OR,* vol. XXXVIII, pt. 1, p. 371.

24. Opdycke's favorite horse "Barney" was shot obliquely in the right shoulder, the bullet penetrating his heart. *To Battle for God and the Right,* p. 199.

25. Confederate losses at Peachtree Creek were estimated at 2,500 to 3,000, while Federal casualties amounted to about 1,600 killed and wounded. Boatner, p. 626.

26. Opdycke was ordered by General Newton on August 1 to build and defend the redoubt, which was to be large enough to hold two artillery pieces. *To Battle for God and the Right,* p. 207.

27. Major General Henry W. Slocum had commanded the 20th Corps

since August 27. Portions of two brigades from the corps' 3rd Division were the first Federal troops to enter Atlanta at 11 a.m. September 2. *OR,* vol. XXXVIII, pt. 2, p. 392-393.

28. Opdycke was given command of the 1st Brigade, 2nd Division, 4th Corps on August 5. The 125th Ohio's transfer to the 1st Brigade occurred September 9. *To Battle for God and the Right*, p. 209, 223.

SEVEN

'Fighting with cold steel and clubbed muskets'

T wo and a half weeks after they reached Atlanta, the 125th Ohio and its division were sent by rail to Chattanooga. The move was a response to General Hood's still dangerous army leaving its camps 25 miles south of Atlanta and striking north to threaten Sherman's supply line. Disruptions occurred at several places, but Hood then marched from northwest Georgia to Alabama, crossed the Tennessee River and in mid-November headed toward Nashville. Most of the Union field forces opposing Hood in middle Tennessee belonged to the 4th and 23rd corps, detached from Sherman's army.

On November 30 the Confederates clashed with these two entrenched corps at Franklin. Several times Hood's men assaulted the Federal works in what Colonel Opdycke termed a "most determined and reckless manner," the "musketry exceed[ing] anything I *ever heard.*" At one point the Union center was punctured. Opdycke's brigade rushed forward to stem the Rebel onslaught in a wild hand-to-hand melee. Shooting continued long after nightfall, when the Federals withdrew and marched to Nashville's defenses. There, December 15-16, Hood's depleted command was decisively routed by a makeshift army under General George H. Thomas.

Just one day after Franklin's desperate fighting, Henry Glenville, a.k.a. "Cato," penned a long letter to the *Cleveland Herald.* He had recovered from his Chickamauga wound, was promoted second lieutenant, rejoined the 125th Ohio in October and at Franklin was in temporary command of Company F. His account of the battle and its prelude was the only one written by a member of the regiment to appear in the *Herald* or *Western Reserve Chronicle.* Neither paper printed anything detailing the 125th's role at Nashville.

Chattanooga, Tenn.,
Oct. 8th, 1864.

Dear Chronicle: One year ago our Regiment was encamped on
the same ground that it has been occupying for the last few days.
We had then fallen back from the bloody field of Chickamauga
and entrenched ourselves as best we could on the brow of a hill,
which position we continued to strengthen until in due time it
became a formidable work and was named Fort Wood in honor of
General Wood, who then commanded our division. It has since
been changed to Fort Creighton, in memory of the commander of
the old Seventh Ohio. Here with an almost empty commissary,
without railroad communication with the base of our supplies,
nothing in fact but a long wagon road leading over the Cumberland mountains, which was almost impassable on account of the
deep mud, with the exultant foe looking down from Mission Ridge
and Lookout mountain, watching our every movement and threatening our only line of supplies, with every thing to discourage and
nothing to encourage. We patiently watched and waited, without
blankets or sufficient clothing and almost without food, through
the cold stormy days of autumn until the invincible Grant came to
our relief and saved an army and with it Chattanooga.

Such was our position one year ago. How can we then but rejoice when we return after nearly a year's absence and find the
then almost deserted town filled with immense warehouses and
workshops, a great bridge over the Tennessee, in fact every thing
so changed that it is hard to realize that this is really the town we
left a little while ago. What a change! The crest of Mission Ridge
is no longer lined with rebel cannon, nor on the lofty Lookout do
we find an insolent foe. Our commissaries are full to overflowing
and the daily trains from the north, with the steamboats on the
river, supply us with all the necessaries of life, besides many of
the luxuries of home. Yes, there is abundant cause to rejoice at
what we see here, but when we look out side and beyond, and see
what the last year has accomplished towards ending this terrible
struggle, we thank God and take courage for the end surely draw-
eth nigh. We have pierced the heart of Georgia, we have wrested
the Gate City from their grasp and we will hold it. Our force is
sufficient to protect the road from material injury until we have
some six months supplies at Atlanta, then Sherman can cut loose
and establish a new base at some point on the southern coast.[1]

This afternoon our regiment marched out to Lee and Gordon's mills on the old field of Chickamauga, to look after Mr. Hood and his rebel soldiers who have come way up here to destroy our railroad and force us out of Atlanta. Thus far it has been a complete failure, and from what we see and learn of the prisoners who were brought in to-day, their case is now desperate.

Colonel Opdycke commands our brigade and has the entire confidence of all with whom he has to do. His temperate habits and soldierly bearing on all occasions do not fail to attract the attention of those who admire such qualities in a soldier. Captain Moses is still at the hospital; his wound gains but slowly. Hopes to start home soon. Lieut. R.C. Rice commands Co. B now, and is well liked by all. Capt. Bates commands the regiment. He is a very worthy officer and deserves a leaf. Captain Powers is Adjutant General on Col. Opdycke's staff.

There have been several changes since we were here one year ago; the *sad* changes are those which *death* has made. We mourn the loss of the noble Burnham of Kinsman, and Dilley of Howland. Others have received wounds which have rendered them unfit for service, and this reminds me that you have taken our old Captain and made him Probate Judge of Trumbull Co. Our Com. polled 39 votes and *all* were Union.[2]

Our Pay-master has arrived and we shall soon be paid.

Yours,   W.

— *Western Reserve Chronicle,* November 2, 1864

Nashville, Tennessee,
December 1st, 1864.

Editors Herald — Nearly two years have gone by since I wrote you particulars of how the 125th Ohio Regiment initiated their fighting career in their maiden skirmish across the Harpeth river at Franklin, Tenn. I told you subsequently how we built a strong fort and named it in honor of our then commander, Major General Gordon Granger, and prophecied it would play an important part in our great tragedy of civil war. Since that time we have marched thousands of miles into the enemy's country. At Chickamauga we held the key to the position on which the noble Thomas, with the remnant of our army, boldly contended for mastery of the field. There we earned the ferocious cognomen of "Ohio Tigers," which was right according to the doctrine which says: "In

"Imminently fitted to command a regiment," Company C's captain Edward P. Bates was in charge of the 125th Ohio from September 24 until early December 1864. His conduct in the battle of Franklin, thought Opdycke, "was almost beyond reach of praise." Bates never was commissioned to field-grade rank, but in August 1865 received major, lieutenant colonel and colonel brevets. Carte de visite by F.L. LeRoy Photographer, Warren, Ohio.

L. M. Strayer Collection

peace, there's nothing so becomes a man as mild behavior and humility; but when the blast of war sounds in our ears, let's stiffen up the sinews and imitate the action of the Tiger."

Under the gallant Phil. Sheridan we planted our colors among the first on the memorable heights of Mission Ridge. At Dandridge, singly and alone, the $1,25 as we are called, checked and baffled Gen. Longstreet's advance. From Tunnel Hill to Atlanta our standard was ever with the van guard of Sherman's column. Under Harker and Opdycke the Third Brigade was ever foremost where danger and greybacks inclined to obstacle the road. First to gain the rocky precipice of Buzzard Roost. Charging at Resaca both these gallant officers were wounded, but we gained the enemy's strong hold. At Kenesaw, as the skirmish line, we led the way for the charging column; we were repulsed, but we carried from the field the form of our dying young chief, General Harker.

Our original colors were torn to very shreds with battle rents, so that our friends recently presented us with a new stand. It was lately on view in your city in all its silken beauty, emblazoned with names of victories in letters of silver.[3] Now, like the former, it is baptized with blood and pierced with bullets. Yesterday at the battle of Franklin it received its first honorable marks. But I am digressing too far. I little thought when I used prophetical language about Fort Granger that we should tread miles of southern soil as conquerers, and return to the starting point and fight one of the fiercest struggles in our war annals.

When Hood's intention had been partially developed, and Sherman's plan ripe for execution, we, the 4th Army Corps, were rapidly sent by rail from Chattanooga to Athens, Alabama.[4] Marched from thence to Pulaski, Tenn., where we were reinforced by the 3d Division, 23d Corps. Fortifying and awaiting still further developments, we encamped. But Hood had been well schooled in Sherman's flanking tactics, and moved on our flanks accordingly.

On the 21st ult. we marched from Pulaski to Lynnville. Left Lynnville at daylight on the 24th, arriving at Columbia to the booming sound of our artillery, who were holding in check Forrest and his intrepid horsemen. Here we were joined by Gen. Ruger's[5] command, the 2d Division of the 23d Corps, which augmented our force to five grand divisions, all under command of Major General Schofield.[6] Splendid earthworks were thrown up on the south side of the town — perfect models of military engineering, for every man from a Major General to a private in our army is a good practical engineer.

The enemy's further manoeuvres necessitated an abandonment of these works, and during the night of the 27th we withdrew to the north bank of Duck river, blowing up the massive stone fort that guarded the town, and destroying the railroad and pontoon bridge as we retired to the opposite bank of that stream. The sight which struck the beholder at the moment our last few pickets crossed the railroad bridge was one of the grandest in the theatre of war. The whole train had been safely crossed. The last of the army moved across just as "The grey-eyed morn smil'd on the frowning night, checkering the eastern clouds with streaks of light."

But a few pickets had now to hurry to the bridge and the match to be applied to the magazine. To Major Coulter,[7] of the 65th Ohio, was assigned the hazardous Guy Fawkes experiment.[8] His combustibles were two tons powder, three tons heavy artillery

L. M. Strayer Collection   Ray Zielin Collection

Brigadier General George D. Wagner (left) commanded the 4th Corps' 2nd Division at Spring Hill and Franklin. His 3rd Brigade commander, Brigadier General Luther P. Bradley (right), suffered a severe arm wound November 29 at Spring Hill.

and 300 boxes ammunition. As the last vidette passed the vicinity of the fort the gallant Major fired the train, mounted his horse and swiftly galloped to our lines. The noise and force of the explosion was immense. This artificial volcano of man could be compared almost to the terrible magnificence of Nature as seen at Etna and Hecla.[9] Rocks and stones hurled and rattled upon the roofs of domestic shelter. The climax of earthly sound seemed passed when the fixed ammunition burst with lightning flashes. The brilliant glare from the burning bridges seemed to contend with the morn for the mastery of light. The grandeur with which we displayed to the enemy the bump of Yankee Destructiveness for a while held them at respectable bay. For two days we guarded the ford with artillery. We had an immense train to send to the rear, which caused some delay.

The advance of our Division under General Wagner[10] reached Spring Hill at noon on the 29th, only barely in time to encounter a heavy force of rebel cavalry, who must have crossed the river several miles below us with the intention of harassing our retreat

and capturing our trains. General Wagner promptly placed Bradley's[11] and Opdycke's Brigades in position. Before dark the enemy were reinforced by his infantry, when they dismounted and attacked the 3d Brigade under Bradley, which had been somewhat
isolated, making three daring and desperate charges. Each time
they were badly repulsed with great loss. Gen. Bradley was himself badly wounded in the left arm.[12]

The remainder of our army, coming up during the night,
pushed up the Pike to Franklin. To Col. Emerson Opdycke with
the 1st Brigade was assigned the duty of the rear guard. Their
cavalry with admirable military alacrity rapidly followed our skirmishers, compelling us to make frequent halts to check them.
Marching into Franklin we found a weak line of breastworks
which had been hastily thrown up a few hours before and occupied by the force which preceded us. The 1st Brigade moved into
a hollow part of the ground, stacked arms, rested and took dinner.
The enemy manifesting no immediate intentions to attack, our
programme of retreat was still being carried out and our train
moved steadily and slowly across the pontoons over the river. The
1st and 3d Divisions of the 4th corps had safely crossed.

It is now necessary to give you some description of the ground
so familiar to us, and on which three Divisions of our army repulsed ten times their number. Franklin, the county seat of Williamson, is on the south bank of Little Harpeth. The stream
winds nearly around the town, holding it as it were, in the lap of
a crescent. The ground between the horns of this crescent and
connecting them is high, and gives a commanding view of the
country on either side; to the north it gradually declines, the
town being on the lowest part — to the south the descent is also
gradual, the country for a half or three-fourths of a mile on the
right and in front open and clear of timber, whilst on the left is a
thin grove set with blue grass. This elevated ground was occupied
by our troops, who constructed a line of rifle pits and made
breastworks — both ends resting on the river. This gave us a line
completely covering the town. On a ridge to the east, and on [the
north] side of the river, is the fort, which enfilades the battlefield.

Hood saw his opportunity, and true to his combative proclivi-
ties availed himself thereof. From a high hill[13] he could easily see
our position, and saw our forces gradually withdrawing to the opposite bank of the Little Harpeth. Military men will not condemn
Hood's generalship in launching heavy assaulting columns, as he

L. M. Strayer Collection

Second Lieutenant Darius W. Payne, Company I, was gouged in the left cheek by a bullet at Franklin. The former Marietta store clerk was promoted first lieutenant of Company H in June 1865.

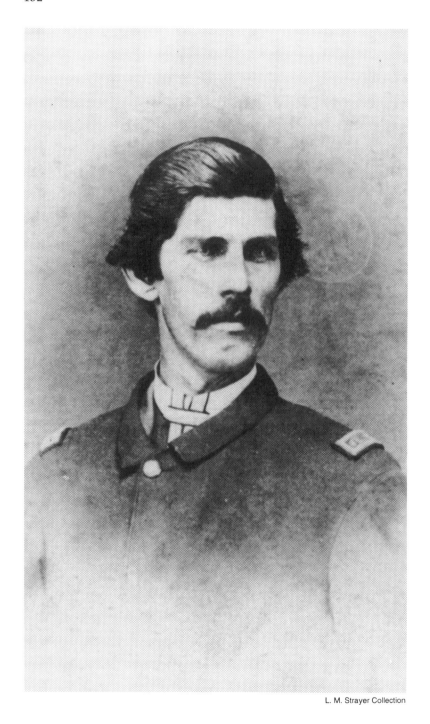

did upon our line. This last hope of recapturing Nashville must at once resolve itself into deathly despair on our reaching the fortifications of the City of Rocks. Had he reached Spring Hill with his infantry a little earlier he would have successfully cut our army in two. But generalship, aided by bravery more allied to fiends than anything human, could avail naught against our devoted ranks. His attack was well managed and hurled with such sudden velocity that it was in itself almost a surprise to us; but we were on our guard.

At about 4:30 P.M. he charged our centre with his infantry massed in columns of three lines. The Brigade which was on this vital part of our line had in it hundreds of new recruits and two whole new regiments. After firing at an angle of about 45 degrees they fled like sheep, panic stricken and dismayed, breaking through our brigade in their haste to escape danger.[14] Great God! The enemy had pierced our centre and occupied our works. What awful issues were involved in this brief second of time. It was but a moment, one of those moments in which the fate of a nation's destiny hangs on but a thread. Was there no relief, no succor? There was! Emerson Opdycke had but to wave his hand and sound the forward when his Brigade was fighting with cold steel and clubbed muskets for the works so basely abandoned by the ignominious flight of its cowardly occupants, moving steadily forward through an avalanche of fleeing conscripts.

Opposite: Standing more than six feet tall, Knox County farmer David K. Blyston was among the 125th Ohio's most resilient soldiers. As a Company F sergeant at Chickamauga he seized one of the regiment's flags after two bearers fell, "unfurled the banner to the breeze and returned it to the guard in a shower of bullets." Shortly afterward Blyston was shot through the right shoulder and walked to Chattanooga for treatment. At Dandridge he was cited for "showing great gallantry" and received a second lieutenancy two months later.

From April to mid-September 1864 he was in charge of his brigade's ambulance train, then rejoined the 125th as first lieutenant of Company A. At Franklin, while overseeing distribution of rifle cartridges near the Carter smokehouse, Blyston was shot in the chest — "the ball having passed close to the heart wounding the left lung." Within three months he again reported for duty (as chief of Opdycke's brigade baggage), although "much prostrated and emaciated." He resigned April 11, 1865. This portrait was made by photographer L.K. Oldroyd at Mount Vernon, Ohio.

The 125th Ohio was led by Capt. E.P. Bates. The 24th Wisconsin by Major McArthur,[15] a chivalrous young officer who received two serious wounds in the breast while endeavoring to rally some skedaddlers. The 88th Illinois, led by Lieut. Col. Smith,[16] combined with the rest of the brigade, the 44th, 73rd, 74th and 36th Illinois Regiments, drove back the enemy's first line and thus filled the gap, and rescued the army from inevitable destruction, wrenching from Hood a victory which had been nearly too cheaply bought.

A moment elapses and a second line bears down upon us; deafening roars of musketry and cannon double shotted with grape and cannister mowed them down by hundreds and knocked to pieces their second line. Hardly had this line been repulsed and slaughtered ere a third one appears through the smoke and seemed to rise from the very earth. They swarm upon our ramparts, bayonets and clubbed muskets again clash in deadly conflict. Rebel officers, hats and swords in hand, lead and incite their followers into our deathly embrace. Rebel color bearers plant their colors on our very works; but it is useless, they cannot pass our invincible line. They charge and recharge on our left and right but with no better result and as fearful loss. Night hovers o'er the struggle but hastens not its close. The deafening roar of musketry and belching artillery is prolonged. Our ammunition is plentiful and the incessant fire we pour through the black pall of night prevents artillery from assailing us and all further formation for charges.

I could enumerate frequent instances of special bravery and amusing incidents in connection with the surrender of a number of Johnnies, but space compels me to shorten up. Our [army's] loss in killed and wounded is very slight compared to the number engaged. We brought 600 of our wounded to Nashville and left about 200 to the tender mercies of the enemy. I estimate our [brigade's] actual loss in killed and wounded to sum up to between one and two hundred.[17] We lost many valuable officers. Captain R.B. Stewart of Co. D, 125th, is supposed to be killed.[18] He was a noble and brave officer, and was one of our best. Lieut. Dave Blystone was shot through the breast, but there are yet hopes of his recovery.[19]

There are numbers of our men who are reported missing who were no doubt killed; as we fought so long and subsequently retreated in darkness we were unable to identify our dead. General

"Arrived at Suburbs of Nashville by A.M.
Men weary & sleepy but glorious. Too
much praise & honor cannot be given
the 1st Brigade. It saved the battle
yesterday. Genl. Cox, Wagner & Stanley
say that Col. Opdycke saved the day."

• Emerson Opdycke diary, December 1, 1864

Stanley was slightly wounded in the neck. Opdycke had a horse
shot under him.[20] Generals Schofield, Cox, Wagner and Ruger
were on the field under the hottest fire and seemed as indifferent
to danger as they would on review.

Let not our friends imagine from the fact that because we left
the battle-ground, and abandoned a number of our wounded, that
we were defeated. No! Daylight might find the enemy far in our
rear and heading us toward Nashville.

I estimate the enemy's loss to be from seven to eight thousand,
killed and wounded.[21] We captured 1300 prisoners and 16 stand of
colors.[22] Prisoners state that Generals Cleburn and Loring were
killed,[23] and General Lee, a nephew of General R.E. Lee, missing.[24]
General Hood is reported to have led a charge in person.[25] It is
very doubtful if he will assault our works here. Speculations and
conjectures as to his present intentions are without end.

I trust your readers will not fail to appreciate our brilliant victory. Had it not been for the 1st Brigade the battle of Franklin
might have been a victory glorious to southern arms. The 1st and
3d divisions of the 4th corps were not engaged. Fort Granger rendered the ford east of the town too hot to allow the enemy to cross,
who put forth great efforts. It also protected our forces when retreating from the town by the heavy and accurate fire it directed
upon the foe.

CATO.[26]

— *Cleveland Herald,* December 8, 1864

Notes to Chapter Seven

1. From Atlanta, Sherman began his "march to the sea" November 15, 1864 with 62,000 officers and men. His forces reached Savannah, Georgia, on December 10 and entered the city 11 days later. Boatner, p. 509, 512.

2. "W" referred to Albert Yeomans, former captain of Company B. Numerous Republican/Union Party candidates were elected to Ohio offices early in October. Long, p. 582.

3. New national and regimental flags reached the 125th at Chattanooga early in October. After viewing them for the first time October 9, Opdycke wrote his wife two days later: "I must say I was disappointed in them, the silk of the Stars & Stripes is inferior, and although the Eagle on the other is very handsome, it is too small for the size of the flag. Crowell will make no more flags for me." *To Battle for God and the Right,* p. 235.

George W. Crowell & Co. of Cleveland manufactured these flags, as it did the first stand of colors issued to the regiment in January 1863.

4. The 125th arrived in Athens at 1 a.m. November 2, and departed at 2 p.m. November 3.

5. Brigadier General Thomas H. Ruger.

6. Major General John M. Schofield.

7. Major Samuel L. Coulter, 64th Ohio, temporarily commanded the 65th Ohio at Franklin. *OR,* vol. XLV, pt. 1, p. 286.

8. In November 1606, Englishman Guy Fawkes was arrested and hanged in London for his role in the so-called Gunpowder Plot to blow up Parliament. *Funk & Wagnalls New Encyclopedia,* vol. 10, p. 114.

9. Etna and Hekla are active volcanoes located in Sicily and Iceland, respectively.

10. Wagner assumed leadership of the 2nd Division, 4th Corps, on September 29. He was relieved of command December 3.

11. Brigadier General Luther P. Bradley commanded the 3rd Brigade, 2nd Division, 4th Corps. He formerly was colonel of the 51st Illinois.

12. After being wounded, Bradley relinquished brigade command to Colonel Joseph Conrad, 15th Missouri.

13. Winstead Hill, about two miles south of Franklin.

14. Glenville may have been referring to the Indiana and Ohio troops of Colonel Silas Strickland's 3rd Brigade, 2nd Division, 23rd Corps, as well as those from Conrad's and Colonel John Q. Lane's brigades of the 4th Corps' 2nd Division.

15. Lieutenant Colonel Arthur MacArthur Jr., father of renowned World War II general Douglas MacArthur.

16. Lieutenant Colonel George W. Smith commanded the consolidated 74th and 88th Illinois.

17. Opdycke reported his brigade's aggregate loss at Franklin as 216. *OR,* vol. XLV, pt. 1, p. 241.

18. Captain Robert B. Stewart, Company D, was killed early in the fight near the Carter House. A native of Fifishire, Scotland, he resided prior to the war at Greenford, Mahoning County. Clark, p. 353; Robert B. Stewart CSR, RG 94, NARA.

19. Company A first lieutenant David K. Blyston's name is habitually misspelled Blystone in the regimental history and other accounts. Service record documents containing his signature verify the former spelling.

20. Opdycke's mount "Ben" was wounded earlier in the day when "a sharp shooter put a ball through [him] about half an inch from my thigh." *To Battle for God and the Right,* p. 249.

21. An estimated 7,000 Confederates were killed, wounded or captured at Franklin. Wiley Sword, *Embrace An Angry Wind* (New York: Harper Collins, 1992), p. 269.

22. Some 702 Confederate prisoners were taken to Nashville with Schofield's army. Opdycke's brigade alone captured 394 prisoners (including 19 officers) and nine battle flags. Sword, p. 269; *OR,* vol. XLV, pt. 1, 241.

23. Cleburne was killed, along with five other Confederate generals. Major General William W. Loring, a division commander in A.P. Stewart's corps, was not killed or wounded.

24. Glenville probably referred to Lieutenant General Stephen Dill Lee, one of Hood's corps commanders, who was not related to Robert E. Lee.

25. Glenville was misinformed. Hood did not personally lead any assaults at Franklin.

26. Writing to his wife December 13 from Nashville, Opdycke professed ignorance of "Cato's" identity and admitted he was "not entirely pleased" with "Cato's" December 1 letter to the *Cleveland Herald.* Opdycke did not elaborate. *To Battle for God and the Right,* p. 257.

1865

Brevet Brigadier General Emerson Opdycke (seated at left) posed with members of his brigade staff at Nashville's Camp Harker in 1865. Among those pictured were three members of the 125th Ohio: First Lieutenant Nyrum Phillips, aide-de-camp (standing far left); First Lieutenant Hezekiah N. Steadman, aide-de-camp (standing center); and Captain Ridgley C. Powers, acting assistant adjutant general (seated center).

EIGHT

'General Opdycke, I have the honor of presenting to you this flag'

After the battle of Nashville, Federal pursuit of Hood's defeated army lasted two weeks. By January 6, 1865 the 4th Corps had reached Huntsville, Alabama, where the 125th Ohio built winter quarters. The 2nd Division was now commanded by Brigadier General Washington L. Elliott, and the 4th Corps by Major General Thomas J. Wood, who granted Opdycke leave to visit Ohio during the month of February. While he was gone, Opdycke's long, valuable service finally was rewarded with a brevet brigadier generalship. He also received a surprise from his old regiment shortly after returning to Huntsville early in March.

Head Quarters 1st Brig 2nd Div 4th A.C.
Near Huntsville, Ala   Feb. 12th 65.

My Dear Colonel: I am glad to write you favorably from the Brigade. Everything is moving with decorum.

The weather which was bad for a few days after you left has cleared off, and the ground is now drying rapidly. Drills and parades will receive vigorous attention this week.

Inspections were held to-day (Sunday) by Regimental Commanders, all which express themselves highly gratified with the result. The Brigade was never better armed and equipped, and the soldiers never took more pride in their personal appearance, than now.

Genl. Elliot [sic] visited us yesterday, and rode through the camps. He was quite delighted, declaring that it was the *best camp in the army*. He intimated that Genl. Stanley would probably visit us soon. We shall not be taken by surprise. Such true soldiers as ours are bound to win praise wherever valor and perseverence can command it.

The sentinel duty is being performed better every day. It will soon be impossible to distinguish recruits from old soldiers.

The Brigade court-martial organized on last Friday. The business before it will probably be completed this week. Captain Barber[1] and Lieut. Griffin[2] of the 88th Ill. who were arrested for drunkenness, are summoned before a Court-Martial at Corps Head Quarters.

There is no indication of a move in the Department. Many officers, including nearly all at Corps Hd. Qrs., have sent for their wives.

Genl. Elliot says we have good evidence that Hood's army has gone after Sherman, with the exception of one Corps and a small body of Cavalry.[3]

By order [of] Genl. Thomas, Col. Blake[4] and a Captain of the 2nd Brigade are to be dishonorably mustered out of the service for persisting in urging their resignations. Having influenced the men who entered the service with them to veteranize, they wish now to leave them to stick it out as best they can.

No business letters have come for you since you left. Four letters have come from Mrs. Opdycke, but as it was no part of her intention that you should read them at home, I will see the spirit of her intention executed.

Col. Stratton came to see you yesterday, and was apparently disappointed to find you gone. From his conversation I think he intends to quit the service soon.[5]

Col. Russell[6] who has been in to see me, says tell the Col. we are getting on finely, and send him my regards.

All the Staff wish to be kindly remembered. My love to Mrs. Opdycke and my best respects to Harmon and Mr. Park. I am

> very Respectfully Your Obt. Servt.
> R.C. Powers, Capt & A.A.A.G.

— Ohio Historical Society

> In Camp near Huntsville, Ala.,
> March 3d, 1865.

Eds. Herald — I don't know of a better way of improving my time on this wet and rainy March day, than in jotting down a few items concerning the 125th, hoping that I may not entirely fail to interest some of the many readers of that very highly prized Union sheet — the *Cleveland Herald.*

Since about the first of November last until a few days ago, by being an inmate of the hospitals, sick and unable to follow the Regiment, I am entitled to none of the honors subsequently attained in the will-be, if not already world-renowned battles of Franklin and Nashville, in which the Union boys so successfully routed, whipped and demoralized Hood's large army that was *going* to take Nashville and then march through Kentucky and Ohio to the shores of Lake Erie. But Hood couldn't quite — at least didn't — match our invaluable and almost immortal Sherman; for *he* accomplished what he set out to do, which was no more nor less than marching his army through the very heart of rebeldom and taking Savannah, one of their very important sea coast cities.

Hood couldn't have been aware of the capabilities of the Union army there assembled around Nashville with the invulnerable Thomas at their head, or he never would have jeopardized his army by placing them in battle array against such a formidable foe.

Our present camp is situated some two miles from the pleasant and interesting town of Huntsville, Ala., on a rolling piece of ground very suitable for a camp and very convenient to those important necessities of an army, wood and water. We repose in the security of a camp guard in addition to the regular line of pickets. As is usually the case in the army plenty of news is afloat after the "grape vine" order, among which, the most prominent, is the occupation of Richmond by Union troops.[7] More anon,

HUGO.

— *Cleveland Herald,* March 14, 1865

Head Quarters, 1st Brigade, 2d Division 4th A.C.,
Huntsville, Ala.,   March 13, 1865.

Eds. Herald — Among the incidents that break in upon the routine of camp life was one which transpired at Head Quarters this afternoon, and will never be forgotten by those who witnessed it. It served to turn the heart in upon itself, and called forth the finest feelings of the soul.

The 125th regiment having been provided with a new stand of colors by their friends in Ohio had, by a unanimous vote, determined to present their old battle-scarred and bullet-riddled flag to General Opdycke, their former Colonel, the man under whose command they had won imperishable honor, and who had so often led them on to glorious victory.

The regiment having been drawn up in front of the General's quarters, he was waited on by a Committee of its officers, and the flag presented to him by Captain R.C. Powers, in the following eloquent and touching language:

General Opdycke: As an evidence of high esteem in which you are held by the officers and men of the 125th Ohio Infantry Volunteers, I have the honor of presenting to you this flag. It is not on account of any temporal value that can be attached to it that it has been deemed a fitting tribute of our respect. As an article of merchandize it changed its value when first it was accepted as a battle flag of the regiment. Since then sacred associations have gathered around it which have endeared it to our hearts, and given to it worth which cannot be estimated in dollars and cents.

It was while burning with resentment at the insult of traitors, who ignored our birth-right of freedom, and ridiculed the principle of equality established by the blood of our fathers, and upon which all our hopes of remaining a united and powerful nation were based, that we first took it up and marched into the field. In many bloody encounters we have met the foe. If we have not always been fully victorious, we have at least never beat an inglorious retreat. The noble sentiment expressed by one of our officers[8] in the midst of the conflict at Chickamauga has been our constant watchword: "The enemy may outnumber and kill us, but he can whip us *never!*" To you for organizing, developing and directing this indomitable courage, is due in a great degree the high honor which we have attained as a regiment.

At Chickamauga, Missionary Ridge, Dandridge, Rocky Face, Resaca, New Hope Church, Kenesaw Mountain, Peach Tree Creek, Atlanta and Jonesboro, this flag has been unfurled in the face of the enemy, and each succeeding time it has come from the field with new laurels and increasing glory.

A thing of beauty, we loved it when first we were marshaled under its bright stars and stripes because it represented the honor and power of our government. Now that its brightness has faded since some of its stars have fallen off, and its stripes become almost obliterated by contact with the foe, it is unsuited longer to be carried into the fight and deserves to be preserved as a sacred relic, to be revered, not so much because it recalls our own sufferings, privations and dangers, but because it is inseparably associated with the memory of our fallen comrades. To that memory the deepest vibrations of our heart's sympathy and love must ever respond.

From Nashville to Chattanooga, from the mountain regions of East Tennessee to the plains of Middle Georgia, their graves are scattered.

Unmarked by any monumental marble, the frowning heights of Mission Ridge and Lookout Mountain, the winding banks of Chickamauga and Chattahoochee rivers, and the rugged cliffs of Rocky Face shall be their eternal tombs.

Attached to this is a list of their names. They number one hundred and eighty-seven brave men — our friends, our comrades, our brothers. They fell at our sides, and we missed them in the fight. On the march their places have been vacant, and in camp and bivouac we have longed for them and wept. In this flag we behold the emblem of the devotion which their blood has sealed.

And to-day these comrades, once so strong, so generous and so brave, but whose voices are now silent, and whose hearts thrill no more with martial ardor, join us in whisperings from the grave in confiding to you its keeping.

Take it; and as long as a stripe or a star remains to reflect the light of heaven, may it be held sacred to the memory of the gallant dead and be a proof of our strong confidence in your ardor, ability and zeal in the defence of our common country, and the cause of universal freedom.

To the new banner that waves over our heads we pledge the same devotion which it was our pride to bestow upon the old one. Although our numbers are few, our ranks much reduced by battle and disease, we are strong in heart, invincible in will, and our confidence in God to defend the right can be diminished by no circumstance of time or place, defeat or victory.

General Opdycke received the flag, and as he gazed upon the tattered emblem his voice could scarcely give utterance to the spontaneous overflow of feeling that filled his heart, as he replied:

Brave Comrades of the 125th Ohio: The poverty of my speech is painful to my heart on this occasion of your unexpected and most touching manifestations of confidence and esteem. Accustomed as we have been for the last few years to sterner scenes, I am somewhat unnerved by your kindness. It moves the heart of your commander, produces a deep feeling of gladness that he has been where men ought to be, where you have been, and leaves the tongue almost dumb for want of eloquence equal to your goodness and your appreciation.

My friend Captain Powers, who has just spoken for you, has seen fit to make pleasing allusions to my career in the National army. I thank you sir for the delightful expressions of this notice and you all for that unyielding confidence and faith which have been a rich reward for the deep concern I have ever felt for your welfare and your honor. Your triumphs under this old battle-torn flag have been signal and conspicuous, not the result of good fortune, but of discipline and valor inspired by a lofty patriotism. My heart has often swelled with pride to behold you bearing this National emblem well to the front midst the thick powder smoke of battle.

On the bloody plains of Chickamauga, when you first unfurled it to the fierce blast of war, your steady fire and the resistless charge of your cold steel caused it to float defiantly for two days against the furious assaults of the enemy; and when the shaft had been severed in two the second time, and the flag cut through with hostile lead, you bore the fragments proudly and triumphantly in the storm. In the memorable and grand assault on Missionary Ridge, you were among the first to sweep the terrible crest and pursue the foe. Dandridge, among the mountains of East Tennessee, you fought almost alone for a day, and won the praise of cavalry Sheridan. When the frowning heights of Rocky Face Mountain were to be scaled, you stormed up its forbidding steep, and planted this banner on its gory crest. You did this alone, and the service was so important the great Sherman gave official praise of it to a whole division. A hundred and ten days under fire during the campaign of Atlanta, you will not cease to remember Resaca, Adairs-ville, Muddy Creek, Jonesboro and Lovejoy, and in connection with heroic and worthy comrades of the Brigade at Franklin and Nashville, you stamped a national renown upon the grim pages of war.

That furious charge at Franklin has rarely been equalled in stern grandeur and importance. It confessedly saved the army from destruction, and added luster to American arms. All through your long campaign for the liberty and nationality of our loved America, this old battle-riven standard has never known dishonor.

You entrust it to me with a roll of those who went down into early and unmarbled graves, fighting under its folds for Country and for God. Precious is the gift and deep my gratitude, and in the years to come, when age shall bow me towards the tomb, I will read this roll with tears for the glorious dead, and embrace this sacred relic with beating heart, and thank God for the good fortune which permitted me to serve with them and you under its thrilling inspirations.[9] Appreciation and praise from those you highly esteem are objects from which the eye of the soldier is seldom turned, but your highest reward, the approbation of your own consciousness, is beyond the gift of your fellows. Their smiles are pleasing incidents, for which alone you would make a single charge on the foe; but when on the fields of carnage and amidst the thunder of battle you have contemplated the starry flag of "true free hearts' only hope," and all the burning glories that cluster around it and spread over the earth beneath it, you have become filled with immortal resolve to *do or die* ....

The obstacles to the achievement of our mightiest destiny are fast disappearing before the streams of patriot blood, and the lights of liberty shine over the land through the wounds of our heroes.

This Mathew Brady studio portrait of Brevet Brigadier General Emerson Opdycke likely was made in mid-July 1865 when he briefly visited Washington and New York City during a 30-day leave of absence. Opdycke was promoted to full rank of brigadier general of volunteers on July 26, 1865.

L. M. Strayer Collection

I thank you beyond the power of expression for the distinguished honor you have conferred upon me, and the hallowed trust you have been pleased to impose.

The tear dimmed eyes of the gallant few that now compose the 125th showed how deeply the words of their loved General affected them. Yes, those men that had so often faced the bristling ranks of the foe, with hearts steeled to deeds of noble daring, wept as they thought of their slain comrades. But the stern, compressed lip showed that the fire of valor they had caught from their former Colonel still burned in their hearts, and that they deserve the name they had won, of "The Tiger Regiment of Ohio."

VICTOR

— *Cleveland Herald,* April 1, 1865

Huntsville, Ala.,
March 16, 1865.

Eds. Herald — Coming to this once fashionable locality immediately after the overthrow and dispersion of Hood's army last
December, we settled down and out of the rude material which
nature so abundantly affords, built substantial log huts 7x10 and
10x10. In order to the better decoration and comfort of the interior of our martial suburban residences, we proceeded with van-
dalic vigor to the demolition of sundry aristocratic mansions, thus
obtaining an ample supply of doors, windows, flooring, &c.

We have lately been reviewed twice — once by the brigade
commander, and once by the Division General. These reviews are
certain prognostications of early activity — certain shadows of
coming events. Soon after, circulars from Headquarters confirmed
our anticipations by warning us to hold ourselves in readiness to
march at a moment's notice. "They say," whose reputation as a
notorious liar still, remains unquestioned, furnishes us with
transportation to Knoxville; there reinforces us with 30,000 additional recruits, plans out a Spring campaign for General Thomas,
who at the head of a mighty host, thunders through the East Tennessee range of mountains, issues out at Lynchburg, joins the little Corporal, thus making the 4th Army Corps present or accounted for at the forthcoming Waterloo conflict.

During our inactivity here this winter, officers and soldiers
have been liberally furloughed. The lucky recipients of these fa-
vors are rapidly returning to their commands; the effects of a
"good time" at home somewhat thinning them down. Brevet Brigadier General E. Opdycke, late Colonel of the 125th Ohio, who has
commanded with honor and success our Brigade since the siege of
Atlanta, has also returned, looking well. A few days ago we presented him as a testimonial of our esteem our battle-torn colors,
which have recently been replaced by a new stand.

The 4th Army Corps have felt much aggrieved at the manner
in which the Paymaster General has overlooked their claims for
adjustment of pay. While the Potomac army and other armies
have been paid to February 28th, we have pay due us since Aug.
31st, 1864. We certainly think we have earned our share, and
while we have been camped here there has been plenty of time
and opportunities to fork over. A soldier's renumeration is small
enough, Heaven knows! and must his family suffer and starve to

enable a batch of Paymasters and their clerks to dawdle and delay over the rolls so that they can collect the *extra pay* accruing to
them when on duty? We cannot reconcile ourselves to any such
absurdity, that Uncle Sam. lacks paper and printing presses. But
amidst all our grievances on this point there is one feature in it
which we view with satisfaction, and that is the blank expression
of despair which has clouded the countenances of the avaricious
horde of Jew sutlers, traders, sharks &c. These greedy cormo-
rants have paid high taxes in the Treasury Department for the
privilege of supplying goods and cheating the army. But alas, for
their sanguine hopes of a rich return, we have no money to fool
away upon them.

CATO.

— *Cleveland Herald,* March 22, 1865

In Camp near Huntsville, Ala.,
March 19th, 1865.

Editors Herald — Seated in the top bunk of our snug little
"shebang" on this beautiful morning, I again take the pleasure of
addressing the readers of the *Herald* and those interested in the
welfare of the 125th O.V.I.

To-day is Sunday, the day recognized and respected by everybody, except the military portion of the world, as a day of rest.
But the soldier, as a general thing, looks upon Sunday as any
other day. He is almost, if not altogether, obliged to respect it not,
and would hardly know when it was Sunday if it wasn't for his
"memoranda" or the never-failing Sunday inspection, which, for
to-day, is already passed.

From our window of four lights of glass, which in all probability helped to compose some Alabama planter's mansion, but which
now lights a soldier's rude shanty, in the distance can be seen a
low range of mountains at the present time undoubtedly the retreat of some roving guerrilla band. In the intervening space lies
spread out to view a beautiful undulating country that in former
years, before that desolating scourge, war, had made its withering
and blighting inroads, boasted of its large cotton fields and tobacco plantations. Across the creek, where the camp guards are pac-
ing their beat on yonder rise of ground, surrounded by old negro
quarters and outhouses, stands a large southern "Mansion House"
(I suppose that's the name to call it), which at present is Brigade

Headquarters, and occupied by General E. Opdycke and staff ....[10]

Yesterday a number of us — 1st Lieutenant Steadman among the number — visited a cave situated in the vicinity of Huntsville, perhaps about two miles from town and a mile from camp. Our entrance was made by descending a ladder of about twelve or fourteen feet in length, minus all but the top and bottom rounds, and clambering down a descent of about fifty feet of broken, sharp and slippery rocks. Perspiring freely, we at last gained the bottom and with our candles, which we had previously lighted, dimly glimmering in the gloomy darkness, we commenced exploring the interior of this rocky cavern of Nature's own construction. After about an hour and a half or two hours, picking our way among loose rocks and enormous boulders, we at last became in a measure satisfied, and again sought the bounteous light of day and the free air of heaven, which we gained in due time, all safe and sound, but puffing equal to a young four-horse engine. But, for the present, I must say "Au revoir."

HUGO.

— *Cleveland Herald,* March 29, 1865

Notes to Chapter Eight

1. Captain Edwin L. Barber, Company A, 88th Illinois, mustered out June 9, 1865. *Illinois AGR,* vol. V, p. 241.

2. First Lieutenant Henry C. Griffin, Company D, 88th Illinois, mustered out June 9, 1865. *Illinois AGR,* vol. V, p. 246.

3. General Hood relinquished Army of Tennessee command in mid-January 1865. Portions of that army, about 5,000 troops, were sent to reinforce Confederate field forces in the Carolinas, where General Sherman then was campaigning. John B. Hood, *Advance and Retreat* (Edison, N.J.: The Blue and Grey Press, 1985), p. 307, 309.

4. Colonel John W. Blake, 40th Indiana, received an honorable discharge in March 1865. W.H.H. Terrell, *Report of the Adjutant General of the State of Indiana 1861-1865,* vol. II (Indianapolis: W.R. Holloway, 1865), p. 392.

5. Lieutenant Colonel Henry G. Stratton, 19th Ohio, mustered out February 13, 1865. *Ohio Roster,* vol. II, p. 639.

6. Lieutenant Colonel John Russell, 44th Illinois.

7. The rumor of Richmond, Virginia's occupation was untrue. Union troops first entered the Confederate capital on April 3. Long, p. 665.

8. Lieutenant Charles T. Clark.

9. The next day Opdycke sent the flag home to Warren under care of Captain Powers, who visited Ohio during a short leave of absence. It remained in Opdycke's possession until his accidental death April 25, 1884. Three years later his widow presented the flag to the 125th O.V.I. Association in Cleveland during its fourth annual reunion. It later was given to the state of Ohio and today is located at the Ohio Historical Center in Columbus. *To Battle for God and the Right,* p. 279; Clark, p. 428.

10. Opdycke's Huntsville headquarters were in the home of former U.S. Senator Jeremiah Clemens (1814-1865), who served as a lieutenant colonel in the Mexican War. *Biographical Directory of the United States Congress 1774-Present.*

NINE

' Those true feelings that actuate
the hearts of Americans '

F ew in the 125th Ohio thought the war could last much longer
when the regiment left Huntsville March 28 and returned to
East Tennessee. The 4th Corps was ordered to block possible
routes of retreat for the two battered eastern Confederate armies
of Robert E. Lee and Joseph Johnston. But the move proved unnecessary. Lee surrendered April 9 at Appomattox, Virginia, followed 17 days later by Johnston near Durham Station, North
Carolina.

Camp near Blue Springs, E. Tenn.,
April 12, 1865.

Editors Herald — On the 28th ult. the 1st Brigade of the 2d
Division, 4th Army Corps, left their pleasant camp at Huntsville,
Ala., for the front, via Bull's Gap, East Tennessee.

The freight cars in which we took passage were none the more
inviting for the peculiar perfumery that pervaded them, having
just been vacated by their previous occupants, which were no
more nor less than a gang of horses or mules. Nevertheless we
managed to make ourselves tolerably comfortable, with the exception of the hours between sunset and sunrise, when — but I will
not attempt to portray those night scenes. Suffice it to say that I
know of nothing that the word "tight" could be more suitably applied to than our situation when in sleeping position. Such a con-
glomeration of legs (let alone other portions of the body) I don't
believe was ever witnessed before. The expressions occasionally
made by some unlucky one that happened to be underneath at
the time, while taking a new position, were amusing if not laugh-
able.

We arrived at Bull's Gap after some hair breadth escapes and a few railroad accidents on the morning of the 1st of April, where we left the cars, marched about a mile up the railroad and went into camp.

We drew clothing and *hard tack* (cartridges) and sent our extra baggage to the rear during the three days we stayed there, and on the 4th of April at about 6 o'clock A.M. we struck tents and took up a line of march (the 125th acting in the capacity of train guard) for Blue Springs, at which place we arrived the same day about noon, having marched some ten miles, and went into camp in regular order with the expectation of staying some time, or at least till the railroad destroyed by Stoneman[1] some time previously is again put in running order, which is rapidly being accomplished, heavy details being made from every regiment in the brigade daily for that purpose, besides the citizen government employees at $2.50 a day.

Last Monday evening the temperature of the army here was raised a number of degrees above zero by an official dispatch stating that Lee had surrendered his army, consisting of 64,000 men, to Grant.[2] It seems to me that it would be almost, if not quite, an impossibility for the same number of men to raise more noise and smoke when peace is declared than was witnessed on the occasion in question. Cheering, shouting and the firing of guns was kept up till after midnight. The smoke and smell of gunpowder everywhere thickly pervaded the atmosphere, and reminded one in many particulars of a battle and battle-field.

It is the general impression among the soldiers that war and secession are nearly played out.

Nature, at least in this section of the country, is rapidly assuming its summer coat of green and variegated colors. Apple and cherry trees are in full bloom.

HUGO.

— *Cleveland Herald,* April 26, 1865

Camp Harker, Nashville, Tenn.,
May 3, 1865.

Eds. Herald — Under circumstances wildly different did we enter the capital of Tennessee a few mornings ago,[3] from those under which we entered the same place five months ago. Then it was in the depth of a Southern winter; we were tired and footsore, having manoeuvered for weeks with an inferior force in front of the army of General Hood, from the Tennessee river until he had violently precipitated his army upon us in that memorable charge at Franklin, now celebrated as the grandest and bloodiest assault in mass of any one of the opposing forces during this war. Our faces were blackened with grime and battle smoke, evidence of that bloody conflict. The massive stone forts that guard the entrance to [Nashville] seemed to frown and scowl an angry welcome as the slow, winding columns of Stanley and Schofield dragged their weary in upon the Franklin pike. The piercing, chilling winds whistled through our very anatomy as we stood burning our shoes at the coals of the camp-fire, and blinding our eyes from its smoke, in order to get a little warmth. Such were the circumstances that attended our entrance to the City of Rocks on that December morning; but mark, how changed the scene.

Spring, ever joyous Spring reigns supreme and clothes with a rich mantle of green the face of nature. The twittering of the feathered tribe makes the air musical with their warbling. The air is fragrant with the perfume of early flowers. The merry sunshine exhilarates the enchanting scene. The black clouds of war are rapidly rolling away. As the last of them recedes from view the Angel of Peace, with all her heavenly attributes, stands before

Opposite: Shortly after breakfast May 14, 1865, a personal note from Brigadier General Washington L. Elliott was delivered to Emerson Opdycke at his Camp Harker headquarters. "The Genl. officers of the 4th Corps," it read, "have been requested by Mr Morse Gallery of the Cumberland 25 Cedar St. to sit in group for a picture. Genl. Stanley has suggested a meeting at 12 to-day with swords & belts, & sashes. I propose leaving my Hd. Qrs. at 11 a.m. & would be glad to have you accompany me in my Hd. Qr. ambulance."

During the short ride to Nashville Elliott and Opdycke learned of the May 10 capture in Georgia of Confederate President Jefferson Davis. "We thought the news ought to cause an unusually amiable expression in our pictures!" quipped Opdycke, who judged the session's exposures by A.S. Morse, Department of the Cumberland's official photographer, to be "quite successful."

Seated from left: Brevet Major General Samuel Beatty, Major General Thomas J. Wood, Major General David S. Stanley, Brevet Major General Nathan Kimball. *Standing from left:* Brigadier General Ferdinand Van Derveer, Elliott, Brigadier General Luther P. Bradley, Opdycke.

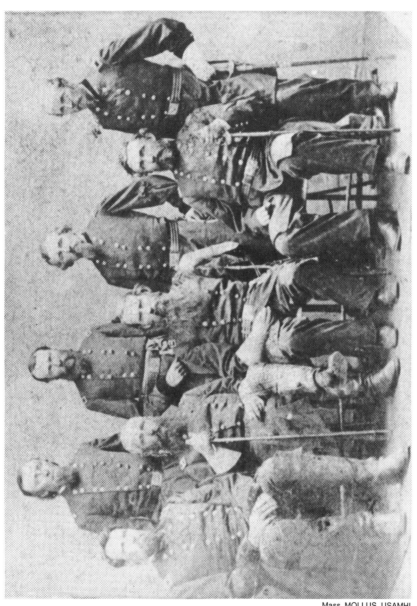

Mass. MOLLUS, USAMHI

us. Those great ugly forts have lost their ghastliness, and now look in their majesty approvingly upon us. Our breastworks of that winter siege extend a useless furrow but still a remnant relic of determined resistance.

Camp Harker[4] (named after the lamented General C.G. Harker, who fell at the head of his brigade while assaulting the enemy's works at Kenesaw Mountain last June) bears all the charms of a stately grove and avenue. The white tents of the army and the symmetry observed in their disposition cannot fail to excite the admiration of those who love the picturesque.

The 125th Ohio Regiment, Lieutenant Colonel Jos. Bruff in command,[5] guarded the train of the 2d Div. 4th A.C. from Blue Springs, East Tenn. to Knoxville, a distance of sixty-seven miles. The regiment never knew a detail of such a nature while an enemy demanded attention in front; never getting, as it is demonstrated in the army, a "soft thing" when something had to be "did" in the eminent deadly breach. The boys did not appreciate the honor as the rest of the command rode through on the cars. Upon arriving at Knoxville we dissected each wagon of which there were over 200 and loaded them upon the cars.

The valley through which we marched from Bulls Gap to Knoxville embraced the towns of Morristown and New Market, on the E. Tenn. & Va. R.R., and is a splendid growing section. The agricultural arts of peace already predominate. New fences lined the route and everything was rapidly recovering [from] the desolation of war. Loyalty so long persecuted in East Tennessee is now proudly triumphant. Guerrillas, cavalry and marauding parties are at a discount. What few infested that region have come in for mercy since hearing of Lee's surrender. No better field for enterprise and capital can be had than that found in the resources of East Tennessee. Let Cincinnati urge and aid the completion of the Railroad via Cumberland Gap to Knoxville and a new and fertile region is opened to her commercial interests. After that Railroad shall be completed I prophecy for Knoxville a population and business equal to Nashville or Louisville.

I have a little to say on the dangers and pleasures of riding on a U.S.M.R.R. and the course observed in engineering them. From Knoxville we rode via Chattanooga, Stevenson and Murfreesboro to Nashville. The safe transit can be fairly estimated as an equivalent to passing unharmed through a severe skirmish — so frequent do accidents occur, seldom happening without loss of life.

So abominably unskillful is most of the track laid. So unstable
are the majority of the trestle bridges on the route that a slight
flood or drift will carry them off entire. But if surprise is awakened at all on this subject it should be why there are not *more
accidents*. One who has passed over any part of the Railroad will
maintain a distinctive recollection of cars in an inverted position,
broken debris of rolling stock, locomotives damaged and their ma-
chinery in a confused mass, burnt ties, bent rails, &c., ornament-
ing the steep embankments on the route. A tie or a rail loosened
or displaced would seldom be seen by the engineer in time to
avert the impending catastrophe. When a train is to be started, it
is divided into from four to six sections, each consisting of a locomotive, caboose and from fifteen to twenty cars. During the journey an interval is preserved between the sections, regardless of
speed or grade of from six hundred to one thousand yards. This
plan would be extremely useful in the event of an attack, as the
troops would be then well in hand.

The passengers on these roads are made principally of that
class who are engaged in seeking a "bubble reputation even at the
cannon's mouth." They carry their refreshments, wardrobe and
defences about their persons. They are not provided with a smoking car or the comforts of a soft velvet seat to be found in an ordinary passenger car, nor are they sheltered from the inclemency of
the weather. They must take the outside accommodations or
none. The inside is reserved for the la—live stock, &c., &c.

As the task of describing an excursion of this pleasant manner
of traveling confronts me, I must express my inability to do the
subject justice. Fancy, ye travellinge publick upe in ye Northe,
yourselves on the curved top of a rickety, broken, dirty old freight
car, jickety picketing over miles of a still more rickety road. I say
fancy, and give your imagination full scope, laying there thinking
as to what premium a Life Insurance Company would allow you
to insure your bones when every jolt of the car and evolution of
the wheels is sufficient to furnish your gravitation elsewhere than
where you are making so painful an effort to maintain it. No
matter in what direction you keep your face, the smoke from the
engine claims a right to miscegenate you, while the cinders have
and exercise an undisputed right to attack your eyes. But though
I would not recommend to tourists this method of traveling, still
it has its claims to superiority. No surly conductor orders you into the soldier's car or the hind car on account of wearing the hon-

ored uniform of your profession. No demand is made to show sundry documents to incur an incision from his pinchers. No boys disturbing your meditations with their chewing gum, &c. Then again there is no second or third class, and no ten minutes for refreshments at any depot.

Numbers of our gray-clad erring brethren, armed with the paroles administered to them by Ulysses, rode through with us from Knoxville on their way to their homes. They were comfortably and well clothed, and the greater part of them very intelligent men. They mixed freely with our soldiers, although there was a peculiar reserve observed in their intercourse. I heard no expression of malice from either side, or taunts thrown out by our men. As a general thing they were hungry and without rations, and our men freely tendered and admitted them to a dividend of their haversacks. They accepted it gratefully, but beyond this exercise of our better nature and their respectful demeanor, I could not fail to see a "cloven foot."

There was a general look of an unsubdued character portrayed upon their features, as if a disposition or resolution to fight again, if ever an opportunity occurred, was gnawing the utmost recesses of their hearts. Their countenance was that of sorrow more than anger. No genuine sorrow for their past treason, but sorrow springing from that deep rooted evil of the human heart: pride.

Opposite: Company B, 125th Ohio, photographed at Camp Harker prior to its mid-June 1865 muster-out.

Seated from left: Orderly Sergeant Rufus Woods, Corporal John Thompson (behind Woods), Corporal Isaiah Brown, Corporal George P. Davis, Captain Ralsa C. Rice, Warren H. Fishel (behind Rice), John C. Mossman, Almon Peck, Corporal Wallace J. Henry, Edwin C. Woodworth, Harvey Giddings, Apollos P. Morse (real name Joel Carr), John W. King,* Patrick Welch, Gilbert L. Cook,* James M. Pollock, John Gillis.

Standing from left: Jesse H. Carey, Sergeant William R. Fitch, Gideon A. Robinson,* James Cranston, Theophile Panquett,* Thomas Brown,* Emory Gilmore, Orasmus Fitch,* Thomas Loutzenhisar, George Pigott,* Norris Meacham, Sergeant Jones K. Warren, Sergeant Albert Mathews.

Asterisks denote men who enlisted between January and April 1864. All others were 1862 enlistees. Identifications were derived from a key written on Emerson Opdycke's personal copy of the albumen photograph.

Mass. MOLLUS, USAMHI

Sorrow that their treason had failed. The most intelligent seemed the greatest maniacs, and were utterly unreasonable in their ideas of future Southern independence. A great many hugged the delusion of an exchange. The phantom Hope seemed to linger by them holding in her hand the prospect of their future success. They acknowledge themselves defeated in a military sense, but they are unwilling to cover more ground. They share the opinions of their leaders. Treason is still with them a virtue. It is infused in their very blood and marrow. They still hate the Nation's flag. They still hate the Union.

We would be wary of even pardoning the rank and file of the rebel army too lightly. The Republic is now well able to vindicate its strength. Forbearance and conciliation have long since ceased to be virtues. If they are not prepared to go full length in sustaining our nation hereafter, let them be impoverished and exiled, and if they are determined to remain as snakes in the grass, let us facilitate the angel of Death to gather them into his folds. Let them give some better evidence than that of doubly damning themselves in profaning their lips with our sacred oath of allegiance, ere that pardon which the magnanimity of the nation alone offers should be conferred. We speak this not in a revengeful spirit, but demand in the name and interest of humanity that the work that has cost us so much blood and suffering be not lightly completed.

Col. Jacques of the 73d Illinois of our brigade and who will be remembered in connection with his visit to Jeff Davis as a Peace Commission[6] has returned to his regiment, and on Sunday afternoon last delivered an eloquent eulogy on our late President. Having a personal acquaintance with him he related many circumstances in his character showing the value of the man for whom we now in common mourn.[7]

A few words on how the army received the exciting news of the "Fall of Richmond," "Lee's Surrender," and the "Assassination." When the news of the "Fall of Richmond" reached us April 3d we were at Bull's Gap, East Tennessee, having arrived there but the day previous from Alabama. Our mission then at the Gap was to encounter Lee in the event of his evacuating Richmond. The receipt of the news did not bring with it, to us, those joyous feelings experienced elsewhere. Nothing, however, was said of which road Lee had taken and we were too well schooled in war to anticipate rather than that in the event of Cavalry Sheridan not heading him off, he would avail himself of the gate we were watching, and either he would be in our hands in a very short time or we would be in his. Then followed the encouraging dispatches of our old division commander (Phil. Sheridan), showing how he was walking into Lee's affections.

We made up in giving vent to our patriotic feelings when the news of Lee's surrender came. I shall never forget that night. Most of the camps had retired; I sat up to gratify my astronomical curiosity in seeing the partial eclipse of the moon, announced in the almanacs for that evening's entertainment, April tenth. An orderly dashed hastily up to Col. Bruff's tent with the dispatch. The Colonel sent word he had a great dispatch to read. Men jumped up from their bunks regardless of the necessity of wearing apparel and rallied round the Colonel, who, standing on a stump near his headquarters, read by the light of a flickering candle the momentous news.

A hushed silence was observed during the reading. When finished three cheers were proposed for General Grant — and such cheers. The moon seemed to stop his performance and stood a wonder-wounded hearer. Then three cheers for Abe Lincoln, and then three cheers for Phil. Sheridan. Of over 12,000 rounds of ammunition which the regiment could account as on hand before the news came, ere morning dawned not a single round had we. Discipline gave way to those true feelings that actuate the hearts of Americans on such occasions. Remember I speak of but one regiment alone, but the same "shindy" occurred in every regiment of the division. The men formed in line and fired off all their cartridges, occasionally by volleys, but generally a mixed *feu de joie*. The different bands turned out and blew what wind they had out of them. Our bass drummer pounded at his drum in blissful ignorance of the necessity for an observance of time. Next morning I saw it in fragments lying out in the ditch near his tent. What little "commissary" was in the command, Headquarters were selfish enough to retain for the special bender demanded on the occasion, and a general sobriety was consequently observed throughout the command. So far had we carried our enthusiasm that a distant command was under arms from midnight till daylight, supposing we were attacked. They were determined they should not be surprised or "gobbled." A Division order the following day prohibited any further manifestations of this character. Standing by a comrade at the time of this demonstration, he remarked to me that he bet the people up in Ohio would go crazy on receiving the good tidings. By your issues of April 10th and 11th I should think you did go crazy, *"you bet."*

The sad news of the President's assassination was received in that same sad spirit that the people heard it, but if a fight was in prospect at the time, the enemy would require a strongly reinforced line where we would attack.

Conclusion. My length demands it. It is not yet certain what disposition is to be made of us, though strongly rumored that we shall ere long participate in a "short, sharp and decisive" campaign in the Lone Star State. Over eight months' pay is due General Thomas' army, still there is no sign of any paymaster to adjust our financial claims as yet.

CATO.

— *Cleveland Herald,* May 10, 1865

Notes to Chapter Nine

1. Major General George Stoneman, District of East Tennessee commander, began railroad-wrecking operations March 20 in East Tennessee and western North Carolina with about 4,000 Federal cavalrymen. His expedition was intended to support Sherman's forces, then near Bentonville, North Carolina. Long, p. 655.

2. In actuality, an estimated 26,765 Confederates of the Army of Northern Virginia were surrendered at Appomattox. Boatner, p. 22.

3. The 125th Ohio left Knoxville April 27 and, traveling by rail via Chattanooga, reached Nashville at daybreak April 29. Clark, p. 384.

4. Camp Harker was located about three miles west-southwest of Nashville off the Charlotte Pike.

5. Bruff was promoted to lieutenant colonel in January 1865. Joseph Bruff CSR.

6. In July 1864, Colonel and minister James F. Jaquess and wealthy New York businessman James R. Gilmore met in Richmond with President Jefferson Davis and Secretary of State Judah Benjamin. Their unofficial visit, an attempt at peace negotiation, produced no results. Long, p. 541.

7. Abraham Lincoln was shot the night of April 14 while watching a play at Ford's Theater in Washington. He died the next morning.

POSTSCRIPT

Cato's letter of May 3, 1865 from Camp Harker concluded 125th Ohio correspondence printed in the *Cleveland Herald*. Like its author, all in the regiment expected to be going home soon. While they waited, General George H. Thomas reviewed the 4th Corps at Nashville May 9. Thomas, observed Opdycke, "was very happy and declared it to be the finest display he had ever seen."[1] As the grand review closed, the general asked for one brigade to form in battle lines, throw out skirmishers and make a final mock charge. Opdycke's was selected. Thomas professed the spectacle to be "first rate."[2]

The regiment remained at Camp Harker six and a half weeks, much of that time hearing unsettling talk of being sent to Texas. Except for Companies A and B, the rumor proved true. The news greatly displeased most officers and men, and prompted a few of the latter to go "missing."

On June 13 the *Herald* published notice that Company A — reduced to two officers, three sergeants, five corporals and 13 privates — arrived in Cleveland the previous night, and all were treated to a hot dinner at Wheeler & Russell's dining hall before going to Camp Cleveland. "Co. A," the newspaper noted, "is discharged under the order mustering out of the service all troops whose terms expire prior to the first of October. The remainder of the regiment [is] under orders to proceed to Texas with the Fourth Corps. The men are all yearning to come home, and the order sending them away is received in anything but a cheerful spirit."[3]

Company B also was allowed to muster out. Having gone to war 30 months earlier at 89 strong, the company under First Lieutenant Henry Glenville[4] numbered just 22 men when it detrained at Cleveland June 23.[5]

Following a 4th Corps reorganization, on June 15 the rest of the 125th Ohio began a long journey south by rail and steamboat. It still belonged to Opdycke's brigade, which now was composed of the 36th and 44th Illinois, 40th and 57th Indiana, and 26th and 125th Ohio.[6] The brigade's ultimate destination was the central Texas coast along the Gulf of Mexico, where it was to form part of an occupation army under Major General Philip Sheridan. Stability was needed in the region, as well as a sizeable Federal force to prevent French political and military meddling in Mexico from spilling across the Texas border.

Enroute, the 125th disembarked near New Orleans, Louisiana, and camped there between June 23 and July 16.[7] Many in the regiment obtained passes to the Crescent City and, among other activities, patronized several different studios to have photographic portraits made. Apparently some continued to question their retention in the army and expressed dissatisfaction. "We have received numerous letters lately from friends of this regiment," the *Herald* reported, "asking why it is not allowed to return home, stating that other regiments who followed it to the field, and others who went at the same time, have returned and are already disbanded .... If faithful and gallant service could be rewarded with a muster-out, the 125th have earned that favor."[8]

In the meantime, Opdycke was placed in temporary command of the 4th Corps' 2nd Division.[9] On July 4 he was granted another leave and traveled to Warren via Washington and New York City. While in Ohio he was elevated to brigadier general of volunteers,[10] a promotion both he and a host of others believed was long overdue.

During his absence Opdycke was kept abreast of events by the 125th's Captain Edward Bates, the 1st Brigade's acting assistant quartermaster. On July 13 Bates wrote that Captain Ridgley Powers (the *Western Reserve Chronicle's* correspondent "Ceylon") had mustered out and left the brigade staff July 8. "The health of the command is declining," opined the quartermaster. "Fevers prevail in several cases. One of our new Orderlies must die." Bates himself was kept quite busy, but "Those who have the least to do complain of its being lonesome. I am more fortunate. We [the staff officers] are all anxious to go to Texas since our minds are made up to go."[11]

After a "very rough and unpleasant" steamer voyage fraught with seasickness,[12] the 125th landed July 23 at Port Lavaca, Texas. At Camp Irwin eight miles inland, the Ohioans found grapes,

226

L. M. Strayer Collection

A Mahoning County native, Captain Freeman Thoman was Company D's last commander. He earlier served as acting adjutant, first lieutenant of Company H and Company A's orderly sergeant.

cattle herds, snakes and mosquitoes in abundance. Temperatures soared "beyond our previous experience," declared one member of the regiment. "Mosquito nets have been issued to us, a new thing in our soldier outfit. No attempt will be made to drill; dress parade in the evening and Sunday inspection will be about the only occasions when we will wear coats or belts. There are to be no pickets or camp guards. No one will want to leave camp in daylight. The heat is too much to face and there is no place worth visiting within ten miles." [13]

Mail delivery was infrequent and verifiable reports at a premium. Excitement in camp was produced by the least tidbit of what passed for news: Captain Steadman's collection jar containing a horned frog, hooded adder, thunder snake, tarantulas and centipedes (August 11); a storm-downed tree at brigade headquarters (August 17); Captain Anthony Vallendar's boil (August 20); mush and milk for supper in lieu of regular fare (August 22).[14] On August 29 a box of marbles made its way to camp, generating "extraordinary zeal" among the soldiers who played games with them like "big boys." [15]

By September 1 it was clear that such a large force in southeast Texas was unnecessary. A War Department order dismantled the 4th Corps, consolidating it into an organization styled the Second Division, Central District of Texas.[16] Opdycke never served with it in the field. After his Ohio leave expired he traveled to New Orleans, where in late August he learned from General Sheridan that he was to be mustered out.[17] Due to Opdycke's urging the same applied to the 125th Ohio's eight companies in Texas. Official word reached the regiment September 17, and the next morning work began on the muster-out rolls. They took four days to complete.[18]

While this transpired, the attention of one regimental officer focused on business far less savory. Captain Freeman Thoman, Company D, was charged with theft and unbecoming conduct that allegedly occurred during the July voyage from New Orleans. At his court-martial September 19-21, Thoman was tried for stealing two engraved, silver-plated spoons from the steamer *Champion's* mess table. The charges mortified the 21-year-old company commander, who bore scars from a hip wound suffered at Kennesaw Mountain,[19] and otherwise was known as an "honest" young man with a "love of fun and practical jokes." The court, however, pronounced guilty verdicts and sentenced the defendant to be dismissed from the service. His long, previously unblemished record seemed forever tainted just on the verge of going home, but Thoman soon breathed a sigh of relief. The court mitigated the sentence in the belief his act "was the result of thoughtlessness, and not of an intent to commit crime." It recommended Thoman instead be honorably mustered out, a measure the Department commander approved.[20]

Thoman and the rest of the 125th departed Texas October 1. Their final trip by sea to New Orleans and river to Cairo, Illinois, consumed 11 days.[21] Three more were required before cars carrying 274 members of the regiment, under Lieutenant Colonel Bruff, rolled into Columbus' train depot. There, reported the *Cleveland Herald,* "they could not restrain their joy any longer, and they made the roof echo with their shouts of gladness."[22]

The history of the 125th Ohio Volunteer Infantry Regiment ended October 18, 1865 at Columbus' Camp Chase barracks. As each man answered to his name, an army paymaster doled out money and a paper discharge. Few loitered around the paymaster's table, most preferring to find quick modes of transportation back to their respective homes. At the city's National Hotel that night, one of them wistfully imagined the feather bed and cooked meals he once again would enjoy. "Good-bye, comrades," he reflected. "The growl of the tiger will be heard no more."[23]

Notes to Postscript

1. *To Battle for God and the Right,* p. 290.
2. *Ibid.,* p. 290-291.
3. *Cleveland Herald,* June 13, 1865.
4. Captain Ralsa C. Rice of Company B had tendered his resignation June 16, and it was approved two days later. Rice, p. 183-184.
5. *Cleveland Herald,* June 24, 1865.
6. Clark, p. 391.
7. Ibid., p. 393, 397.
8. *Cleveland Herald,* July 21, 1865.
9. *To Battle for God and the Right,* p. 301.
10. Opdycke's promotion occurred July 26, 1865. Warner, p. 349.
11. Bates to Opdycke, July 13, 1865, Emerson Opdycke Papers, OHS.
12. Bates to Opdycke, July 22, 1865, Emerson Opdycke Papers, OHS.
13. Clark, p. 398, 400.
14. Ibid., p. 404, 405.
15. Ibid., p. 407.
16. Ibid., p. 407.
17. *To Battle for God and the Right,* p. 305.
18. Clark, p. 409, 410.
19. Freeman Thoman CSR, RG 94, NARA.
20. File #oo1416, Court-Martial Case Files 1809-1894, Records of the Office of the Judge Advocate General - Army, RG 153, NARA.
21. Clark, p. 416-419.
22. *Cleveland Herald,* October 20, 1865.
23. Clark, p. 421.

APPENDIX

'How I yearn to have done with this sort of soldiering!'

A myriad of problems beset 125th Ohio recruitment in the fall of 1862, and difficulties raising enough men to complete the organization only seemed to worsen in 1863. For four months after Colonel Emerson Opdycke left the Buckeye State January 3, 1863 with the regiment's first eight companies (A through H), a mere handful of recruits was all that could be scraped together for Companies I and K. Attempting to remedy the situation, Opdycke placed his trust in a 24-year-old Methodist minister from Athens County, Ohio.

David Hastings Moore had been captain of Company A, 87th Ohio. When that regiment's detachment of reenlistees led by Henry Banning was assimilated into the 125th early in December 1862, Moore briefly was considered to fill the lieutenant colonel's slot. Instead, the position went to Banning. Moore, after turning down the 125th's chaplaincy, "retired" and returned to his wife, Julia, and young son at Athens. Home life and ministerial duties were interrupted in late April 1863 by an offer emanating from the Statehouse in Columbus. As phrased in the regimental history, "The officers have signed and forwarded to the Governor of Ohio a request for the appointment of David H. Moore as Major, he to recruit the two additional companies required to make a full regiment."[1] Opdycke personally had appealed to Moore in March, and followed up his entreaty with another early in May.

Moore accepted. Appointed May 22, he replaced Major George L. Wood who resigned a month earlier due to physical disability.[2] Moore initially believed there were "very good prospects of success" in recruiting for the 125th, and with enthusiastic ardor began correspondence with Opdycke in June about the effort. The major quickly realized, however, that his new job would be anything but easy. For one thing, thousands of eligible Ohioans already had volunteered and were serving in the army. In the summer of 1863, Moore reminisced decades later, "The draft had been declared and ... important elections were pending in Ohio. Intense opposition to the war had developed. Sympathy with Secession was open and defiant. Oath-bound organizations were armed and drilled. Treasonable badges were flaunted. In some places children were forbidden by the school authorities to wear Union colors. Recruiting was most difficult. Companies were harder to raise than regiments had been. Meetings for volunteers were molested; men inclined to enlist were subjected to protest and intimidation." [3]

In spite of such setbacks, Moore persevered. From June through the end of November he kept Opdycke regularly informed in minute detail of both failure and success. Occasionally his letters (part of the Emerson Opdycke Papers at the Ohio Historical Society) dealt with subjects other than recruiting. One penned in the wake of John Hunt Morgan's Ohio raid provides a thrilling description of the major's experiences during the Confederate cavalry foray that skirted his hometown. It is published here for the first time, along with the bulk of Moore's letters written to Opdycke while he labored to raise Companies I and K. They not only illustrate how perplexing the task became, but also shed illuminating light on Civil War intra-regimental activity that seldom is encountered in print.

The major "is a charming correspondent," Opdycke mused July 12, 1863 at Hillsboro, Tennessee. "I wish him here." [4] Five weeks later Moore was promoted to lieutenant colonel. [5] Near the end of the following month Company I was mustered at Ohio's Camp Dennison, and finally, in early December, enough men reported at Columbus to be mustered as Company K. [6] With the latter in tow, the jubilant recruiter left immediately for East Tennessee, joining the 125th at Blain's Crossroads in mid-January 1864. One year had passed since the rest of the regiment first embarked for the front.

Opposite: David H. Moore posed at Cadwallader & Tappen's Marietta, Ohio studio in 1862 while captain of Company A, 87th Ohio. As lieutenant colonel, he commanded the 125th Ohio in the field from January 14, 1864 until his resignation the following September.

Ken C. Turner Collection

David Moore commanded the 125th during much of 1864's winter, spring and summer. He and Opdycke developed a smooth working relationship and close friendship. The "gallant little parson," thought Opdycke, was " a fine officer and a Charming good man."[7] Conversely, in Moore's estimation the colonel was his "brother-father."[8] Not until the fall of Atlanta was their field service severed. Declining health was something Moore no longer could ignore when he tendered his resignation September 10. Officially, however, his stated reason was "that I may resume preaching the Gospel. I am urged to this by the *most solemn convictions* of *duty,* which I *dare not* disregard."[9]

Endorsing the request, Opdycke wrote in tribute to his friend: "Col Moore is a brave & capable officer. I think it doubtful if any better officer of his rank is serving in the armies of the U.S. He is as conscientious as he is gallant and chivalrous and it is only profound respect for individual convictions of duty that prevents me from disapproving this tender of resignation."[10] On September 24 Opdycke confided privately that when Moore departed for home that morning, "he seemed very much affected .... tears came into his eyes and he thanked God he had been permitted to associate with me .... I regret deeply to lose him."[11]

Moore did follow his convictions once he laid down his sword and returned to preaching. Until his death late in 1915, the ex-recruiter resided variously in Ohio, Colorado, China, Oregon and Indiana, where he held both ecumenical and secular offices.[12] In 1900 he was elected a bishop in the United Methodist Church.

Athens, O., June 25, 1863

My Dear Colonel: Your *very welcome* letter of 19th reached me last night.

The man who wouldn't love you is incapable of love. I live on your letters, nearly — with just a *little* bread & butter thrown in.

I see from the contents of yours, that my two, written since the call for 6 mos. men, had not yet reached you. I suppose they have turned up 'ere this; so I need not repeat ....

I have lost Lt. Aleshire[13] — from whom I expected much. However, he may not succeed in his efforts, and bite my hook yet. It has spoiled part, at least, of my plans for this county; and has damaged 3 years' volunteering generally ... and rendered a "change of base" necessary. I have now ready to work whenever opportunity offers — Lt. Gage[14] at McArthur, Vinton Co., O.; Lt. Jno. V. Stewart,[15] Coal Run, Washington Co.; Lt. Welch,[16] Camp Chase — all of whom are mustered in as recruiters; Capt. Spaulding,[17] under certain contingencies, at Ravenna; Lt. Hall, Portage Co., upon whom I do not at all depend; and am about completing arrangements with Lt. Manchester,[18] a live man & successful recruiter, at Ashtabula; also with Mr. Hoyt,[19] the Commissary sergt. of the old 87th, of Ashtabula Co. I've dropped our "Lt." C.R. Waters. He's of no force, in my estimation. If I can get Manchester & Hoyt in rightly, they will work up their county in the conscription *well*. Through Welch we will work up Knox, Richland, &c. It seems pretty certain that nothing worth repeating can be accomplished until the conscription. Preparations for that are being rapidly completed. Work is retarded somewhat by Copperheads, in certain sections.

To-morrow I start for your section, agreeably to your advice. Shall spend Sabbath in Columbus, looking after Welch, and attending to your private business with the Gov. & Adjt. General. The latter I do hope to be able to work up for you. Monday, I shall go to Cleveland, where I will remain until I see Manchester, if possible. Thence to Warren, where through the influence and aid of your friend, Mr. Park, I hope to get some plan for your congressional district in operation. If necessary, I will preach them a sermon. I shall be sure to see your family, & to try to get acquainted with the people.

I hope the detail you have asked for may be granted. They could do much good. If Vallendar[20] goes home, he certainly can make some arrangement for recruiting, at least among the Germans. Capt. Parks[21] can do much good. He is a splendid recruiter.

I am greatly obliged for the account of the regiment's experience. The preference shown the 125th was certainly complimentary and highly gratifying, but it does not surprise me in the least. I knew you would make it a *model*. And what bores me is, that while all of you are improving in the drill right along, I am so engrossed in general affairs as to make no particular progress; and more than all, that I am not being schooled by you in regimental reports, &c.

I share your enthusiasm and your hopes, and appreciate the necessity of the regiment being filled, and that by nothing more remote than the conscription. I do not seek to spare myself any labor, nor any expense that I can stand. I hate to be answering the questions continually — "When are you going to your regiment?" — "O! I thought you were in the service?" &c. It plagues me ....

Milroy[22] has been whipped terribly at Winchester — *cut all to pieces.*

Hoping to hear from you often, I remain, yours Eternally for God & my Country, & the 125th!

D.H. Moore

Columbus, O., Saturday, June 27, 1863

My Dear Colonel: I have just returned from an interview with the Governor and Adjt. General. The Governor thinks the conscription will be enforced next month. *He assured me he would fill us up from it.* And this without any hesitation .... He also stated that it would be impossible to date the muster of the officers back — as you & Gen. Hill[23] had talked; so all your hard work is "gone up." While in the Adjt. Genl's office I learned that your detail was granted. The men will have to report to Col. —— (I can't just now think of his name) & I must use them through him.

I am, of course, perfectly excited over your change of programme, and look every minute to hear of your being in a fight. Let me have full particulars, and let the regiment know that it is not my will that keeps me from the field. I leave you to see to it that they don't get prejudiced before they know me.

I shall spend the Fourth in your place, or wherever Trumbull County celebrates it. May Heaven watch over and protect you!

Earnestly yours,
D.H. Moore

N.B. Welch is doing quite well. He has recruited 9 from the six months' men. One of his [detachment] will go to Cleveland to try it there. The Governor is now raising a regiment of *state troops,* very unfortunately for recruiters ....

Cleveland, July 6, 1863

Dear Colonel: I have this moment returned from Warren, where I have been since Thursday noon last .... I had consider-

Clark, *Opdycke Tigers*   Clark, *Opdycke Tigers*

Second Lieutenant Horace Welch   Captain Anthony Vallendar

able of a *public* airing you will admit when I tell you that I made a war speech and preached twice, once in the Methodist Church, once in the Baptist. Don't know whether I made much of a baulk or not, others must decide that. I shall feel now acquainted with Trumbull Co., and can work up their conscripts well. You and the 125th stand very high, and the drafted men will very generally prefer our regiment. The Military Committee and Provost Marshal will do all they can for us.

They are raising no 6 months' men of consequence. A few are volunteering in Heavy Artillery. The great majority will stand their chances in the conscription.

I saw your old Father [24] and heard him pray. It was a feast of religion. Your little boy [25] was full of life. He is as smart and affectionate as he can be. So far as family is concerned, we are alike; each bound by a wife and boy.

I will write you from Columbus to-morrow, when I hope to be able to give you the results of our operations among the six months' men. I have had two active men, Hanson's, here for a week. They will effect something, if anybody, almost can. However, if they fail, I have one more plan — Manchester will try it. He is energetic and *persistent*. All I am afraid of is, that the Adjt.

Genl. will not allow appointments to recruiting officers for the vacancies in the old Companies. If he will not, it will injure my plans materially. I will try hard, however.

We are rejoiced at the success of our arms under Rosecrans and Meade. Of Rosecrans we have no fear; but tremble for the Potomac Army even in the hour of its greatest triumph ....[26]

I leave at 2 p.m. for Columbus. Have just learned that Hanson[27] has returned to Camp Chase, from which I infer he could accomplish nothing of importance here. Welch will be mustered as a detachment as soon as he has 42 men. Will see him before I write to-morrow.

Most Earnestly Yours,
D.H. Moore

N.B. The indications are still that the conscription will come off this month, near the 20th perhaps. The enrollment of the 1st class is not yet complete in your District; but the Marshal has had orders to begin to prepare the ballots. *That* looks like work.

Columbus, O., July 7 '63

Dear Colonel: Arrived here last night. It is now 8 A.M. Have been out to camp this morning. Lt. Welch is absent recruiting in Tuscarawas County. All his men are to report to-morrow. They expect to be mustered this week. I saw Sergt. Hanson. He reports that Col. Senter[28] absolutely forbid his recruiting unmustered 6 months' men. Outside, he saw no show; and hence, reported back here. I am satisfied that it is almost impossible to effect any thing in Cleveland. Artillery & the Navy take all; and that *all* is made up chiefly of *strips* of *men,* such as would do us no earthly good. Lt. Welch's men are largely sprinkled with *boys.* They look *hard*-y enough, & may be regular *pine-knots* for what I know. He is more particular now, and will take none but the very best. He — Welch — is a capital fellow — full of the hang-on principle, and plucky to the bitter end. Experience will rub off his faults and leave him a fine officer.

So far I have secured good men for recruiting except one, in whom I am afraid I have been sold. However, I am not prepared to speak positively just yet. He don't seem like your style of man. I shall have to act upon the principle of "a bird in hand is worth two in the bush." Capt. Spaulding remains undecided, meanwhile the golden opportunity is passing away. I *must* have Manchester of Austinburg, Ashtabula [County] appointed; & shall wait on the Adjutant General to expect it as soon as ten o'clock comes on.

You cannot imagine, my dear Colonel, the joy of the people over our eastern & Western victories. You have outgeneraled, scattered, & demoralized Bragg & freed Tennessee; the Army of the Potomac seems to have routed and ruined Lee, while old Dix[29] and his veterans are making Richmond tremble. Glory to God in the highest!

I will finish after my interview with Genl. Hill. Start for Athens, via Cincinnati, at noon.

(Later) — All right: Manchester will be appt.

A man by the name of Hess is recruiting for us at Dayton. He reports to Adjt. Genl. 7 men. I will try to hear from him before I write again. He stands well here.

Truly & Eternally,
D.H. Moore

Athens, O., Friday, July 10, 1863

My Dear Colonel: It seems a long time since I heard from you last; but I know that, since you have been on the move, writing and mail facilities must have been scarce. What a march you have had! What roads! The rains that were swelling the Potomac to the terror of Lee were the good angels of Bragg. But your march has been triumphant in spite of storms. The forward movement to Tullahoma must be reckoned among the most brilliant passages in the history of the immortal Rosecrans.

Now that Vicksburgh is ours, & Grant's army free to co-operate with you, more triumphs must be ours. God grant that they may speedily come!

Meade's victory at Gettysburg was magnificent, but not decisive. The great struggle must yet take place. The indications now are that Lee will make a desperate stand on the old Antietam grounds. We are concentrating an overwhelming force against him; &, if, instead of pushing on to Richmond, Dix comes up on the other side of the Potomac, effectually cutting off Lee's retreat, the hopes of the Rebellion must receive a death-blow.

It is amusing to hear people talk — "I reckon there will be no draft now," "What glorious victories! no need of conscripts now!" "If these victories prove as decisive as represented, there will be no conscription, do you think there will?" Even Gov. Tod is reported to have said that the next thing he expected to hear was *peace!* It is evident that we have the Rebellion on the ground; let us have 300,000 new men to kick it to death! They are enforcing conscription in Rhode Island, Pittsburgh, & will soon, I believe, be at it in Ohio. I expect to leave for *"Head Quarters 125th"* about the middle or last of August with a splendid lot of conscripts. "There's many a slip" &c, and I am no exception to mankind at large.

Since writing last, I think I have struck one good lead at least. I wrote 1st Lt. W.T. Johnson, Co. B, old 87th — the Adjutant knows him well. He is a lawyer in Ashland, O., and assistant provost marshal of his district. His reply is *warm* and *encouraging.* He desires to re-enter the service, and concludes his letter, "Let me hear from you again, and, if possible, let me become identified with the 125th O.V.I." I wrote him immediately, and shall await his reply with considerable anxiety.

I calculate on 11 [commissioned] places; six in the new companies, and five in the old. If a good man is *instrumental* in securing 30 conscripts, I will be at liberty to assure him a lieutenancy, will I not? At any rate, I have been going upon that supposition, and do not see how I am to *"make certain"* of our needed number otherwise.

I have not heard from Capts. Vallendar & Parks, &c, and do not know whether they have reached the state. I do not see what is to hinder V. from being very useful in Cleveland.

Your wife sent mine your photograph. The picture works wonders. My wife is more willing than ever for me to enter the service. My parents think, perhaps, I was not so enthusiastic after all; and my neighbors with one accord say, "He is a *splendid* looking man!" Candidly, my family could have received nothing which they would prize more highly or from which they would derive more satisfaction. They feel acquainted with my Colonel; and it don't seem to them so much like my going among strangers.

There is a dearth of news just now from my men. Indeed, we can expect nothing until the conscription. May the Lord hasten *that* in his own good time! Regards to all.

Eternally Yours, D.H. Moore

Jno. Morgan has invaded Indiana. If he gets out, somebody's to blame, *sure.*

Athens, O., July 27, 1863

My Dear Colonel: Your three letters of 6th, 12th, & 19th are at hand — the two last came together in the mail of Saturday.

For the first time since my commission with your regiment, I have failed to comply with your orders, as to writing you weekly. The reason will, I trust, be abundantly satisfactory — John Morgan has been raiding through our county, & I have been after him. Athens County is as dear to me as Trumbull is to you. Perhaps I had better begin with one week ago last Tuesday, July 14.

Morgan had reached our [rail]road below Loveland, & was moving toward Hillsboro. The militia of southern Ohio had been called out. Seven suspicious characters — supposed to be Morgan's men — had been arrested at Chillicothe, with tickets for Warren, 6 miles below Athens, where there are very many large trestles. It was feared the design was to break our road by burning one or more of these. The Road asked for a guard. I volunteered to take care of three trestles; raised a squad, repaired to the scene, but had no foes to encounter except mosquitoes. Returning Wednesday morning, I found my valise ready packed at the depot, and Surgeon McHenry's dispatch requesting an interview in Columbus that day. So on I went. Everywhere one could see the militia flocking in to the various rendezvous. The whole country seemed alive. Reached Columbus at 9:30 p.m. City full. Found Dr. McHenry's register at the Buckeye,[30] but he and all the [recruiting] detachment, except Parks, had gone. Was with him until he left, the next — Thursday — afternoon. They had reported to the Adjutant General & been assigned as follows: Surgeon Henry McHenry, to the counties of Henry, Fulton, Defiance & Lucas. Sergt Nyrum Phillips to Trumbull, Mahoning, Ashtabula & Geauga. Reports to McHenry. Capt. S.B. Parks — Knox & Wayne. Sergt. J.A. Bell[31] — Delaware, Morrow, Marion, Richland. Reports to Parks. Capt. A. Vallendar — Cuyahoga, Lorain & Erie. Sergt. Major S.A. Smith — Ashtabula, Mahoning, Trumbull & Portage. Reports to Vallendar. I wrote them, but, as yet, have received no answers. When I returned I found [a] telegram from McHenry, saying that he could not wait, but would write. This detail will be a tower of strength to our cause.

Returning Friday, our train passed through Hamden. There had been skirmishing with Morgan within 6 miles of there, and a fight was expected. Sent my valise home by a friend, got off the train, and volunteered as pro tempore high-private in Co. B, 2nd Regt. Ross Co. M[ilitia]. Lieut. Gage — one of our recruiting officers — acting adjutant, discovered me, reported, and soon came with an order to report myself as an aide to Gen. Gillmore.[32] The only improvement was in the *horse*. We had not proceeded far before we learned that Morgan had flanked and run away from Runkle's[33] forces. Counter march was the order, & the command was embarked on the cars for Athens. Arrived there. I said "good-by" to the General, saddled "Jo Hooker," and at 12 at night, started to lead a cavalry company to the Athens Militia already long on their way to intercept the marauder.

Found them drawn up about 15 miles from Athens. Fed my horse; and about half past four A.M. Saturday, started on an independent scout, in company with two young men, one unarmed, the other carrying a defective fowling piece. Proceeding some five miles, we ascertained that the Rebels were en-route for the Ohio [River] at Pomeroy, & would pass through Rutland on their way.

"Rutland!" is the word. Arrived within a mile of the town, dispensed with the unarmed [man], and go on, followed by the shot-gun, formerly a Color Corporal in the 87th, Jo Pickering.[34] Every man takes me for one of Morgan's troopers. How obsequious he is! Meet what I take to be a resigned Federal officer. He makes the same mistake. Scare him nearly crazy, much to the amusement of *"Shot-gun."* Go on. Outskirts of town are reached. The rebel advance, some 500 strong, has passed through, the main body not yet come up; town held by the stragglers of the vanguard. O, for [a] half dozen cavalry! We could have some prisoners!

We advanced cautiously until we were opposite the church. The Rebs were in the main part of the town at the cross-roads; so we were comparatively obscured by [a] turn in the road, and safe. Against my will "Shot-Gun" dismounted, concealed his horse, gun, spurs, & every thing of a military character, & being dressed in citizen's clothes, sauntered down into the town. I was dressed in military, my horse was equipped, and I dare advance no further. Soon the citizens came running to me, urging me to escape, as the main body was just at hand. "Shot-Gun," too, came with the same report — "The road is just black with them!" I tried to persuade him to mount & be ready to act with me, but he wanted to see more of them, & so, went back. Just then a grey-back came dashing round the corner on a splendid sorrel horse, right plump at me. I was then the quickest man, the best man. I drew on him instantly. He whirled his horse broadside, & threw himself alongside Indian fashion. Away screamed my ball, &, waiting only to see him whirl & put spurs to his horse and dash back toward the main body, I gave "Jo Hooker" the word, and sped faster than the winds beyond their reach. From "Shot-Gun" I have since learned that some fifty chased me, but soon gave it up; & that my ball just broke the skin along the back of the Rebel's neck. I enjoyed it. Twas refreshing.

Our Militia returned that day to Athens. Morgan, repulsed at Pomeroy, was moving toward Buffington's Island. I waited for our pursuing Cavalry, but hearing nothing from them returned that night home.

The next morning, equipping a fresh horse, I started again. This time, for Tupper's Plains, Meigs county. Timber had been felled in every road. Heard about 10 miles out that there had been a fight that morning at the Island, & that Morgan had lost many prisoners.[35] I had for a companion Charlie Ballard,[36] a leading merchant of Athens, armed with a *Colt.* Finally reached sight of the contending forces within a mile of the Plains. Our men were drawn up in line of battle; in the distance we could see the dust rising where the Rebel column was moving, apparently toward the Ohio at Hockingport. It was nearly dark. Met a gentleman who formerly lived in those parts & who advised us, if we wished to join our Cavalry in the pursuit, to turn back and take the cross-road to Coolville.

We expected to meet our forces where the cross-road leads into the Coolville road, supposing they were moving down the Coolville road. As we neared the place ... we halted, and our friend went ahead to reconnoitre. By this time it was dark. We could hear men talking and hear horses moving. Our guide reported all right and we went on. A column of horsemen was moving down the cross-road toward us. We deemed it prudent to fall back beyond a gulley. I waited on the rise beyond to ascertain, if possible, the character of the approaching party, all the while well nigh positive they were our men. Thought I saw men slipping along the skirt of woods to my rear. On came the column. The advance halted below me, dismounted, drew on me, and ordered me to halt and dismount. Challenges passed as to who [the] other side was. They claimed to be Union men; I to be "all right." Suspecting them, & yet so confidently expecting to meet our men right there,

I was more than perplexed. O for a single moment of light!

"Dismount and advance!" comes the second time. Still I parley. The third time: "Dismount and advance!" I knew what follows further non-compliance; so reluctantly enough my right leg slips off the saddle, and leading my horse, I advance. Doubt doesn't last long. Pistols are presented to my head, and the ever horrid word crashes through my ears, "Surrender!" One devil dives for my finely caprisoned horse, two others bear off my fine navies, while a third presents his pistol to my breast and demands my purse. I demur, and demand to be reported to their general. Another interferes in my behalf and the highwayman retires in disgrace. One of the dismounted men, alluded to before, by this time comes up, & patting me on the shoulder, exclaims, "That's the time we were too sharp for you, old fellow!" The rest of the party are soon brought in. Can see the fires blazing brightly where our men are sleeping, while Morgan turns their flank and rides unmolested away!

After plodding along on foot, with guards on either side, for a while, I began to realize that the Major of the 125th was a prisoner and in a pretty plight generally! To get away then became the question. I was clothed partly as an officer & partly as a citizen. Had a citizen's coat and cap. Darkness had prevented their noticing my pants and vest. My horse was rigged with equipment I had received but a few days before. The fellow who got him came back to me and [asked] — "Who *are* you, anyway?" "O, nobody but one of the ragged militia." "Thunder!" responded he, "I thought I had a Colonel!" It was well for me that he was so easily satisfied.

I made myself very sociable, telling them it was hard to be a prisoner, but it would be all right soon; that they were trapped beyond redemption; that thousands of men guarded every road and every crossing — very big stories for the truth. Found out all I could. Found the men expecting to be taken somewhere, yet thinking Morgan would get them away if any man could. [They] admitted that there was no military point to the raid, & that it was a mistake to cross the Ohio. I found that my captors were of Cluke's[37] Ky. Cavalry, Lt. Col. [Cicero] Coleman[38] commanding, with whom — baring the circumstances — I formed a very pleasant acquaintance.

Finally, after a march of some eight miles, the column halted at a cross-roads, some 17 miles from Athens. I seperated [sic] myself a few paces from the rest of the prisoners. Their guard was changed, and they were marched into a house to be locked up. I was not missed. Carelessly sauntering among the rebels, I finally got behind a horse, slipped over a fence, fell prostrate into the high grass beyond, and did some of the tallest crawling on record. Found I was on the wrong side of camp & would have to make a half circuit of it, in order to reach the Athens road, avoiding pickets on four roads. I will not weary you with details of how I got to the woods, ran, walked, and crawled as occasion required and the nature of the ground, all the time in sight and hearing of the camp, until the road was reached. At the second house, I got a horse. It was then 5 o'clock Monday morning. By 9 I was home again. Your Major was *free,* but minus a splendid horse and equipments, worth at the lowest $225.

The Rebels ran square around our men and broke for the river below Pomeroy at Cheshire. Resting until evening, mounted once more on "Jo Hooker," with a squad of good men, start again for the front. 15 miles out, rested for the night. Off early next morning. By ten o'clock hear of a fight at Cheshire, Morgan losing another thousand or so. Push on. Pass the battle field. Meet several of my friends who tell me they have found my horse. Look at him and agree. Visit Gen. Shackleford [39] and obtain order for horse and equipment. Col. Coutts,[40] 2 O.V.C., under a prior permit, has selected that horse, & now gets the Genl. to countermand my order. I can only have my horse by securing him one just as fresh. To this end I visit the prisoners' camp, but find no fresh horses.

Accidentally run across Cluke's men and Col. Coleman. Make myself known. Disclose my rank. Enjoy his surprise, and laugh with him over my escape. He asks if I have any pistols. I tell him his men got my horse, pistols, and all. The officers have the disposal of their private property, and he immediately turns over to me his horse, equipment, & revolvers. I reciprocate by presenting him a souvenir in the shape of a flask of whiskey and a roll of greenbacks. We part *friends* at once and *foes.*[41]

But my horse — what of him? To cut it short, I bought a fresh horse for $100 and ransomed him. The horse I reclaimed is a splendid one, but the joke is still on me — I brought home the *wrong horse!*

Morgan with the remnant of his band whirled again, crossed the Muskingum [River] at Eaglesport, and struck the Ohio yesterday at a point 9 miles above Wheeling, but was repulsed, and is now moving on up toward Pittsburgh.[42]

The question is, havn't I had quite an exciting campaign, with *home* as a base of supplies? Our work has not suffered by this interruption. Every thing in Ohio gave place to the Morgan excitement.

* * * * *

I will now take up your letters and answer your questions *seriatim.* My appointment bears date May 22nd. I am more than obliged for the interest you have taken in getting the muster in shape.

Hall is at home and expects to stay there, unless he secures his commission. He has no idea of going back to the regiment as a private. Is he not a deserter and liable to arrest as such?

How will I work it to secure commissions for those who earn them? Anticipate no difficulty so far as the new companies are concerned, but will the Governor commission lieutenants for the old companies upon my request? Will I not have to await recommendations from you? I will not fail to notify you as soon as the battle is over how we stand. Will attend to the case of Lt. Stuart the first time I am in Columbus. Shall be very careful about appointments. No man shall have one, *unless he earns it.* When the time comes I shall use every endeavor and influence I can control to favor our cause. Will secure the Springfield arm, if possible.

I am glad that you are now on a war-footing with the paymaster, and will secure you a certified copy of the order at my earliest opportunity.

We certainly expect to elect Brough[43] by an overwhelming majority. Morgan's raid has made him many votes.

I come now to yours of 19th. Think your estimate of the number of men the draft in Ohio will put in the service in all probability correct. Shall do all in my power to secure our proportion, & more too, *if I can.* Have heard nothing from Spaulding lately, and have left him entirely out of the account.

Agree with you perfectly as to the unfitness of Welch for Captain now. He has worked so persistently and under such discouraging circumstances, that now the brighter hours approach, I would like to see him rewarded. I have all along thought it possible to satisfy him with the 1st lieutenancy, and think it may be so arranged yet.

If it can be done and not violate my word or injure us in men, I will work *Moses* and *Whitesides* in. The sergeants will have the best chances in the world to help themselves. I will look to Smith and Phillips, and will not forget the "Chronicle." [44] In one word, my dear Colonel, I will do all in my power to carry out your wishes to the letter, having, however, as the inflexible rule — *"Men are the great object; every thing to give way before it."*

The Governor is more sensitive on the subject of six-months' men than almost any one else. He will not allow of any recruiting whatever. Think it an evidence of the Governor's gullability that he continues to repose confidence in such a scamp as Senter. He is a notorious rascal. I am slightly acquainted with Brough, and will make the 125th all right.

I already am *proud* of our gallant little regiment. Shall strive by honorable conduct to be worthy a place in its ranks. Am glad you are under Crittenden.[45] He stands high in these parts as a gentleman and a fighter. And what a blessed thing it is to be connected with the ever-victorious army of Rosecrans! — an army that has never learned how to be whipped!

Manchester writes me that he has already secured 7, and confidently expects to have 20 volunteers before the draft. Think the conscription will speedily be enforced as soon as things settle after Morgan's raid. O, that it was already here! I long to get the men and join you in the field.

Remember me to "Ad." Sorry to learn of the failure of Capt. Baugh.[46] Had hoped better things of him. Hold on to him till after conscription. He has great influence in Knox. Can't you make any thing out of him? Is there no military in him? Will write again this week.

Ever Yours
D.H. Moore

Athens, O., July 29 — Wednesday A.M.

My Dear Colonel: Yours of 20 & 21 came by the last night's mail, and I hasten to answer.

One of two things is certain, either you have been misinformed as to the occurrence of the draft in Ohio, or I have been lost in a *Ripvanwinklian* sleep. Sure it is, that I know nothing as yet of the enforcement of the conscription. I have made use of inquiries in every direction as to how the drafted men will be disposed of — whether by assignment or permitting them to elect their regiments, but have been able to arrive at nothing definite. The provost marshal of your district, as well as the marshal of ours, up to this time, has no *official* idea as to how it will be managed. The prevailing impression is, that they will elect their regiments; and upon this I have been relying, making my arrangements accordingly. If the assignment is arbitrary, I shall claim our two companies and our proportion for the regiment in addition. This will be fair, and I think I can make the Governor "see it." Nor will I let him off with *words*. My "modesty," I know, will be my chiefest foe, but what of modesty when *men* are at stake? I will take the cry from you — "Pitch in!" Shall stick out for 550 men, but dear me! Well, "rosin the beau" — "Jordan is a hard road" &c.

The requests I will attend to carefully. Rest assured as soon as the work begins — and I am very impatient for it to open up — I shall devote every hour and every moment to the securement of our great need. All I can do now is to get *fully ready*. I would not be taken by surprise should the draft begin to-morrow. I am ready and waiting. Still, if the men are not permitted to choose for themselves, I will be very greatly disappointed.

You remember my speaking about getting a son of our provost marshal, a member of the gallant old 36th Ohio, to recruit for us, if possible? Well, he was here to-day to see me. He is ready to take hold. I have written for his appointment. If men choose regiments, the influence he commands will win every time. I feel quite elated over the acquisition. Now, going upon our former understanding, counting Moses & Whitesides as captains, am I right as to the vacancies? —

Co. A	1	2nd Lt.
Co. B	1	1st & 1 2nd Lt.
Co. C	1	2nd Lt.
Co. D	1	2nd Lt.
Co. H	1	1st & 1 2nd Lt.
Co. I	1	1st & 1 2nd Lt.
Co. K	1	1st & 1 2nd Lt.

1 1st Lt. for Adjutant, 1 Sergt. Major. Twelve commissions — 5 first lieutenancies, 7 second. I want to be exact, so as to keep my word faithfully.

Having, I trust, a proper conception of our need of men, I shall to the extent of my influence supply the want. I am very thankful for the suggestions and words of counsel you have so kindly given, and trust you will continue to afford me the advantage of your judgment upon every subject that will at all affect the great issue before us.

Eternally Yours,
D.H. Moore

Athens, O., Aug. 2, 1863 Sunday p.m.

My Dear Colonel: As I start for Columbus in the morning, I must drop you a line to-night.

I have made a few changes in our recruiting arrangements. The appointments now stand thus.

Lt. Horace Welch, Lt. Chase — head-quarters Camp Chase.

Lt. W.A. Gage, McArthur; Lt. Sterling Manchester, Austinburg, Ashtabula Co.; Lt. Jos. Hess — from whom I can get no answer to my letters — Dayton, O. Have written for appt. for Henry Barber,[47] son of our Provost-Marshal, Harman, O. Also for appt. for Louis Amberg, Chillicothe. Have not yet received answer from Columbus.

The time of Lt. Hanson & Stewart, having expired, I have ordered the first — he is not our Sergt. Hanson — to be reversed under *no circumstances;* the latter, not without personal application or new recommendation from me. So you see I number 7 only; ten with the three sergts. detached; 11 counting Sergt. Hanson recruiting for Welch's Co. All this upon the supposition that the conscripts will select their regiments. If this proves to be true, I will get three more to work. Some of them will fall short — enough to save the Sergeants positions, as you desire. It is better to have too many men than too few. If any appointee fails of 30, of course, he can expect no commission.

If I succeed in arranging for Whitesides & Moses, as I have but little doubt of being able to do, then, if the "Chronicle" can not fill the bill for the adjutancy, I shall suggest Lt. J.H. Jenkins,[48] of my old company, for your consideration. He is an exception in every way, better than an A No. 1 man, if possible. I submit his case to Whitesides. Let him testify. However, it is by no means certain he could be induced to re-enter the service.

So far as order & penmanship are concerned & energy, Manchester would pass splendidly. He might lack *voice.* Gage is also a tip-top officer — just a little *"frilly"* (that isn't in Webster — ask Whitesides); he is a fine penman, *quick, ready, brave,* well-drilled, experienced, & I have no doubt would be delighted with the position. I *know* Manchester would. But I go according to the text — the "Chronicle" *first.*

If the conscripts have this privilege, & any are taken from our county, I have struck, to secure them, what seems to me a splendid deal, for 30 at least. On the other hand, if they are assigned, then I shall pursue the plan you suggest in every feature. And how the thing is to be done, I shall find out right off. I more than suspect that it will be *this week* or *next.* Our Provost thinks so.

I will note your "requests" up to date, that you may add or modify if need be:

Date of May 15 — Desire me to make arrangements for permanent recruiting in Ohio, mainly at Cleveland. (This I think must be accomplished through a detail from Capt. Vallendar's Co. I know of no other plan).

July 12 — To hear as soon as I know result of campaign in officers and men, so that you may arrange full complement of officers for regt. (How near can I come to you by telegraph?) Also, to see that Lt. Stuart [Stewart] is made Capt. Co. D, vice Spaulding, resigned. Need over 500 men. Springfield gun. Certified copy of your pay-order.

July 19 — Sergt. Major & 1st Sergt. Co. C to be commissioned if they deserve it, & to help them all I can. Moses to be Capt. Co. I; Whitesides Co. K, if possible. To enlist a good man for Sergt. Major. To give "Chronicle" first show for adjutancy.

July 21 — Reports of Classes. 10 Mahan's Outposts. 10 Mahan's Field Fortifications.[49] 1 pr. pants [and] 1 pr. boots of F.A. Brown, tailor, Bank St., Cleveland.

I have to laugh heartily whenever I think of the "sly hit" in one of your last: "We are drilling a little, *so that you won't be ashamed of us when you come"!!!!* Alas, me! it is well that the position of Major is *"ornamental."* Don't bore me too much. I'll catch up, if hard study under your most excellent tuition will do it. All my spare time I devote to tactics now; but it is but little like *practice in the field.*

If conscripts choose, I shall get an energetic, worthy, influential preacher to work among them, with a view to the chaplaincy, as per instructions. Will write you from Columbus.

Most faithfully yours,
D.H. Moore

Athens, O., "Bright & early"
Thursday morning, Aug. 6, 1863

My Dear Colonel: Granting that we get the full number
assigned us — 440 — we will not fall far short of 500 men, perhaps exceed it. Welch has 36, Manchester 9, Hess 12 = 57. Manchester will in all probability get 20. Gage & Amberg 15, Hess 15,
the detail, say 20, but I'm afraid that's large, say 10. That would
give us, with Welch's men, 96 men, which with 440 = 536. But full
allowance must be made for uncertainties and disappointments.

I am using every effort to push our recruiting appointees to
most constant activity just now. We ought by all means have
Gage & Manchester in the regiment. They are *superior.* Gage will
not be surpassed by one officer in 500 for drill & snap. Enquire of
Col. Banning, the Adjutant, or any one acquainted with him *in the
field.*

Manchester's men have among them some of just the kind we
want. He is energetic, quite well educated, and is a *tip-top* fellow.
If he secures twenty good and acceptable volunteers, I have promised him the first lieutenancy of Whitesides' company. I hope this
will meet with your approval. He merits the position by his ac-
quirements and the character of his recruits; & I cannot ask him
to take a lower one. Every recruit, in view of the mode of dispos-
ing of the conscripts, has almost a double value.

What can we do for Gage? He was forced by Judge Plyley to
stay away when you promised him the position of orderly. He now
says he *will* go, position or no position, as lieutenant, adjutant,
sergeant major, or high private, *with a musket.* Of one thing I am
confident — he would at once take rank with the very first of our
officers. He has his faults, to be sure, but his excellencies obscure
them. I do not think we ought to let slip the chance to secure him
to us. Such men do not grow on every bush.

Suppose Hess gets 15 men and sticks at that number; what
shall I do for him? You see this makes no provision for those to be
placed in the old companies, leaving those chances to the sergeants who really deserve promotion, and leaves positions open in
the new companies for Smith & Phillips. *Please* write me fully on
these points. I will write ... to [the] Military Com. of Trumbull to
secure their action to get their conscripts placed in our quota; also to each recruiting officer to urge him to greatest activity.

Ever Yours, D.H. Moore

Athens, O., Monday morning,
Aug. 10, 1863

Dear Colonel: Yours of Aug. 2 reached me Saturday evening, and gave great pleasure. I suppose you have received my telegram & letter from Columbus, also letter from Athens, all sent last week?

The detail I telegraphed for must be made before we can touch a man of the conscripts. I told Genl. Mason[50] that we already had a recruiting detail in the state, and asked him if they would not do for this purpose. He said, "Not without a modification of their order and an addition to their number. The War Department orders that three commissioned officers and six privates shall be detailed from each regiment to report to the commandant of the nearest rendezvous of conscripts to the place where it was organized. Your detail must be ordered to report to me, at Camp Chase." Hence, you see, it is of the utmost importance that this detail be promptly made and sent on. I write thus fully upon this point now, for fear my telegram and letters were not sufficiently explicit ....

A letter received from Dr. McHenry the past week informs me that he finds it very difficult to recruit men — every thing seeming to have settled down for the draft. He had secured *two* & hoped to be able to get a few more. I have not heard from any of the rest except Smith, who reports great apathy upon the subject of enlisting. Have written to all, save Phillips; told Dr. Mc to write him — urging them by every consideration to do their *utmost* to secure volunteers, as every man thus secured is worth double to us at this time what he ever was before.

Hope and *expect* to realize nearly a hundred volunteers by the time I am ready to leave for the field. But men are slippery, and our plans often fail and disappoint.

Have had another application for the chaplaincy of our regiment. Rev. Jas. Mitchell is a *good* man, & would move through a regiment respected and beloved by all, but his health is quite poor, and, I think, would not serve him as the chaplain of the 125th. I would like very greatly to have our regiment *well furnished* in this department. I know I shall need every influence and appliance to enable me to lead a religious life; and, God being my judge, I want to lead no other! I have a minister in my eye who would suit you and serve the regiment to advantage. He is a

ONE MORE CHANCE

TO ENLIST IN THE

125th REGIMENT!

This favorite regiment is now at Hillsboro, Tenn, and is in the corps of Gen. Crittenden, one of the most popular commanders in Rosecran's grand army.

COL. OPDYCKE

is regarded with increased favor by his men, and the 125th stands second to no other regiment in discipline and good order.

Pay, Advance and Bounty, same as formerly.

The undersigned may be found for a short time at the NEW YORK STORE, Warren, Ohio, ready to receive recruits for the above regiment.

S. A. SMITH,

July 22, 1863-tf Recruiting Officer.

This 125th Ohio recruiting notice, replete with two typographical errors, appeared in the *Western Reserve Chronicle* September 16, 1863.

self-made man, a man of the people, well versed in human nature, a first-rate preacher, a sharp and smart man, and a man of great and unsurpassed energy and devoted piety — Rev. C.C. Felton. Circumstances may render it possible for us to secure him ....

Athens, O., Aug. 13, 1863

My Dear Colonel: Yours of 7th inst. came to hand this evening. How shall I find language to express my thanks for your so *great* kindness? I *cannot!* May God help me by actions to prove my *gratitude* and deserve your continued love!

I receive and shall present your recommendation of my promotion, not without many and serious misgivings of my ability properly to discharge the responsibilities of the new position, yet determined, if it lies in me, to give you no reason to regret the step taken, and to make myself, by diligent application and unsparing zeal, worthy [of] the great confidence you have reposed in me. For the regimental order accept my cordial thanks ....

So far from letting the arbitrary assignment hinder, interfere with, or retard my exertions to secure recruits, I have made it the reason for redoubled diligence, urging our recruiting officers to use every effort to [achieve] success. They stand thus now —

Welch	36
Hess	16
	4
Manchester	13
Dr McHenry	2
	——
	71

Have not heard from Capt. Vallendar or Capt. Parks.

Received a letter yesterday from Sergt. Smith. He seemed pretty blue. Said he thought it would be impossible to recruit a single man at Warren. I answered him, urging him to keep a good heart and work on, that a recruit was worth double to us now what he was at any former period ....

To return to your letter. You say — "Consider yourself authorized to take our assignment, and take the first drafted for fear they won't hold out." I asked Genl. Mason whether I and the detail already here would not be sufficient to take charge of our conscripts. He said that unless my order and theirs was for that specific purpose, he could not turn our quota over to us. As to taking the first drafted men, the War Dept. not only fixes the order of precedence as to armies, but in each army of the regiments; so we will have to "bide our time." Say we get the 440 for which we are booked, and we now have as per statement, 71 recruits, making an aggregate of 511. And we hope to add 40 to the recruits, making a total of 551.

Think you are correct as to the adjutancy. Your estimate of Lt. Jenkins is reliable. He is one of *"nature's noblemen."* Have written him & will again; conditionally, however, upon the "Chronicle."

Whoever suits you for Chaplain will, I am confident, please me. The qualities I deem essential are dignity, common sense, piety, and energy, with, at least, ordinary oratorical powers. You need never expect to find any *"cool shade"* in my friendship for you, chaplain or no chaplain, Won't you mind *that?*

As soon as I receive the letters from you, which I am positive are on the way, I will see the Governor and have a full and satisfactory talk, as you suggest, & will report promptly. Have no doubt of his cheerful concurrence with your views as to concert of action between himself and his Colonels. They are so just that he cannot avoid it. Would that I was with you now. This waiting for the conscription is *so* wearisome! But I am content, believing like you, that "all's *well* that *ends* well."

> Ever Faithfully,
> D.H. Moore

Athens, O., Aug. 24, 1863

My Dear Colonel: I had finished my work in Columbus, and was at Loveland on my way here, Thursday last, when a telegram from Father sent me back.

I found yours of 12 inst. awaiting, having been forwarded from Athens. From the contents I reasoned thus — Moses & Whitesides must be commissioned the same day with Powers & Stinger, & many of the other promotions are conditional upon those captaincies; they can't be made till we realize our quota from the conscription & organize "I" and "K." Hence, these recommendations require attention *just before* leaving for the field, or, as your letter has it, "so that I may go to you with my pockets full of commissions."

I was somewhat puzzled to know what to do about the immediate commissioning of Capt. Bates. He being the only commissioned officer now with his company, I have thought best to await your definite answer upon this point. I can get his majorship any day — Tod has already approved his recommendation. I only wait to learn if you wish it issued *now* or not, until I secure the rest ....[51]

In confidence. What do you think of Hamblin[52] — formerly captain in your old regiment? I have had intimations indirectly that you do not regard him favorably. Tell me exactly. I have entertained a very high opinion of him *as a military man.* He is unquestionably [and] unusually well-drilled in the school of the company, movements of the battalion, & evolutions of the line. Nash's[53] men think every thing of him; and when I supposed the

Enlisting August 4, 1863, Company I recruit Charles Miller was wounded
at Missionary Ridge just nine days after joining the regiment at Chattanooga.
The 19-year-old private, who apparently enjoyed smoking, posed for army
photographers Greenwall & Stringham and mustered out in September 1865.

conscripts would choose their regiments & we would only secure whom we influenced, then I wrote Hamblin as to the captaincy of "I." He was right in. But as soon as I learned the disposal of conscripts and your wishes, I immediately wrote him the former fact, withdrew every encouragement I had given, and told him I could do nothing for him except [if] he brought recruits. He writes me from Columbus that he is bound to go, though it be as private in Co. I. If Welch secures men enough to muster two lieutenants, I presume Hamblin will be elected 2nd lieutenant & so mustered. Now write me fully — your *will* is my *law.*

Suppose such should be the case, then Smith would have to be provided for elsewhere, and that would lead to a modification of the entire programme. Again: the Washington dispatches yesterday stated that only 12000 would be drafted under this call from Ohio! I confess that the thought makes me sick, but although discrediting it, to some extent at least, we must provide for such a contingency. That being the case, say we get 100 conscripts. We now have 82 recruits. Shall I organize the two new companies, making them minimum? It seems to me this would be the better way, securing, as it would, our regimental muster. But I submit the question. Again — if this is so, Gage must be nothing higher than sergt. major, & the adjutancy reserved as promotion to some[one] who would otherwise be disappointed. Hence, my silence to Jenkins. Please send me recommendations providing for this contingency, which may God grant never may arise!

Mr. Parks writes that anything the Trumbull Co. Mil. Com. can do to place their conscripts in our regiment will be done at the proper time.

I have been looking over a little volume entitled "School of the Guides." It meets a decided want. Have our guides studied it? Would it be best to bring ten copies?

Am watching news of your progress with intense anxiety. May God shield you and our glorious little regiment!

Ever Faithfully,
D.H. Moore

Athens, O., Friday Night, Aug. 28, 1863

My Dear Colonel, How I yearn to have done with this sort of soldiering! I do not want to complain, but when once clear of this scrape, never say *"detached duty"* to me once! You can hardly form any conception of the reluctance men have to volunteer, & of the influences brought to bear to prevent their so doing. It is next to impossible to effect any thing at all. The recruits obtained for 'I' have cost more time, anxiety, & *labor* than the raising of an entire regiment at certain periods. It was easier then to enlist *twenty men* than *one* now.

Hess is doing better than any of the rest. His appt. reads "Captain," hence his title. He is full of energy. Has recruited *hundreds of men* since the war began, but has uniformly been swindled out of a place. I send you his letter received tonight with the enclosed recommendation, in order that you may judge for yourself.

The men, so far as I have seen them, are average. Welch has some few boys — tough, though. Manchester's men are superior [in] every way, many of them having been in the 87th, several being school teachers — one having been commissary sergt. 87th, another, whose letter I send, adjutant's clerk.

Your remarks on the situation and *what we need* command my feelings and judgment. They persuade and convict at once. You must be careful, or I shall begin to think you ought to rank Stanley[54] and with the cavalry of the Cumberland spread terror & dismay through the very heart of the confederacy. Your plans comprehend such rapidity of movement that I almost half suspect your genius is adapted to Cavalry, & that you only stick to infantry for the same reason that many men do to agriculture, because it is the foundation of prosperity & the backbone of strength. Am I right or wrong?

I will attend to the charger; by no neglect of mine shall you ever be compelled to exclaim, "A horse! a horse! my Kingdom for a horse!"

Would to Fortune I were with you now, in this glad eve of your certain victory. I await your answer to my later letters with great anxiety.

Ever Faithfully,
D.H. Moore

Columbus, Sept. 17, 1863

My Dear Colonel: Night before last late, I heard of a detachment of sharp-shooters at Camp Dennison which was about to be assigned somewhere into another branch of the service. I "went for 'em." Early yesterday morning I saw Capt. White, my right

hand man in Adj. Genl's. office, and asked why they could not be
given to me. "They are already under promise to the 9th Cavalry"
— said in a manner that made me think — perhaps, if the offi-
cials in that office were as anxious to see us filled as they are to
favor other regiments, they could have done it long ago. I then
went to the Governor. He was crusty, & gave me to understand
that they were to go to the 9th Cavalry. I shifted and tried him
for a chance at the Governor's Guards. He answered at once &
positively in the negative, & forbade any argument on the question.

I then got transportation to Camp Dennison, & started.
Reached there about 3:30 p.m. Saw Lieut. Coonrod[55] & his men.
Every thing looked blue. The men "couldn't see it" for infantry.
Twenty three had signed a petition to be assigned to the 9th Cavalry. The first lieutenancy was promised Coonrod. I made friends
with a few influential ones, made Coonrod all right by promising
him the Captaincy of "I" — as much as I wanted it for Moses.
They had been in camp since June & hadn't been mustered or
received a cents pay. Finally, I got them together & made them a
speech. I praised them for remaining steadfast under so many
discouragements, assured them it was of such stuff good soldiers
were made. I then descanted on the many disadvantages of an
independent organization such as they would have been had their
company been filled; "set up" the fine qualities of the 125th, told
them I had the men to fill up their company, & could get it mustered the next week so they could receive their bounty, &c., and be
soldiers of the finest regiment in Wood's splendid division of Rosecrans' unrivalled "Army of the Cumberland."

I promised the best shots the privilege of skirmishing as sharp
shooters the first fight we were in, ate supper with them, and, to
my utter surprise, of their own accord 29 out of 34 signed a petition to be transferred to the 125th! Only three refused, & they
are the disorderly. I then took the Captain ... & went to Columbus .... Got here at 4 o'clock this morning, completely heading off
the Major of the Cavalry. As soon as we could we saw the Governor. He couldn't resist, but the thing stuck in the Adj. Genl's
office. We went in and talked the matter over, and the result was
an order signed by "David Tod, Governor," assigning Lt. Coonrod
& his detachment to us!

In my last I believe I spoke of *Providence;* this, I speak sincerely & reverently, is the *fruit.* I hope to be able to arrange it for

Sergt. Smith to be 2nd Lt. of this company. I will send him & Bell, Parks, & all the stragglers I can collect on with "I," which I hope to start next week.

Coonrod has seen nearly 18 mos' service, first in the 3 months 14th, Col. Steedman,[56] then first lieutenant in [the] 48th, Col. Sullivan;[57] appears military, & has a good reputation. He is from Defiance. His scholarship, I judge, is not very good ....

Ever Faithfully,
D.H. Moore

N.B. — There are two other matters I ought to state. First, Coonrod was in Shiloh & Corinth. Second, there is a *rumor* that [the] War Dept. has ordered 3 new infantry regts. from Ohio, & that $302 will be paid bounty to every recruit for them, $412 to veterans, & that this (302) will only be paid to them & the 12th Cavalry. If that ungodly discrimination is made against old regiments, I shall at once withdraw all my recruiting officers and ask to be ordered home to you. It is so unfair that I cannot credit it, as yet.

Columbus, O., Sept. 22, 1863

Dear Glorious Colonel! I knew it would be so, I knew it could be not otherwise. I knew you would lead the 125th to immortal honor. The dispatch that passed through to your wife only confirmed my faith. Tried, tried by fire, and found glorious! Thank God! But, and my soul sickens, you *lost heavily.* My poor soldier brothers! mangled and bleeding. May the God of consolation be near you to assuage your pain, inspire you with fortitude, and prepare you for *all* the future!

This suspence [*sic*] is terrible. *Who's* gone? I fear to hear the answer. But you're *safe!* Thank God! My *brother-father* is spared to me! And I was not there! I *will* not murmur.[58]

How acceptable the new company will be. I go to Dennison tomorrow to organize it, & hope to embark it for Louisville Friday night ....[59]

Faithfully, D.H. Moore

Athens, Sunday Night, Oct. 4, 1863

My Dear Victorious Colonel: Yours of Sept. 23 reached me Saturday last. Grieved at the loss of so many brave men, I am nevertheless glad and proud of the record made by the 125th. What a glorious hour it was for you when General Rosecrans personally thanked you and your command for their conspicuous gallantry! Thank God for it! Thank God for your safety! But I tremble to think of the risk you ran — "he remained on his horse all day," exposed to *storms* of shot and shell. What would have become of us if you had fallen? I shudder at the bare thought! May God in his providence shelter you still is the ardent and constant prayer of my heart ....

I appreciate your intention when you say, "Am glad you were not here. You would have been killed." But I esteem it my greatest misfortune not to have shared with my brothers their *first fight.* The only thing that in any degree consoles me is the tolerable success I have had recruiting, and the thought that the reinforcement will be doubly grateful at this time. I earnestly desired to accompany the men forwarded, but feared to leave without your consent or order; the permission granted in your last is a week later than the time the company started. I am at loss what course to pursue, what course will best promote our mutual interest. Would that I could see you to consult! Whether I ought to avail myself of your permission and gratify my own inclination by at once going, or whether I ought to stay to give personal attention and direction to our recruiting efforts under the stimulus given to volunteering by the great bounties now offered — $302 & $402 — to decide this puzzles and perplexes me. Our losses in the late engagement make it doubly important to secure as many recruits as possible. I feel quite confident that I could have another maximum company by the first of November. It seems impossible for my plans to fail. Look at the "ropes."

Moses, Powers, Smith, Rice, Phillips &c in Trumbull & region.

Vallendar & Sergt. Seydler,[60] of Co. I, in Cleveland.

Hamblin in Geauga, & Manchester in Ashtabula.

Hess & Heikes[61] in Dayton, Sergt. Hatfield [62] in Defiance.

Krug in Scioto, Barber in Harman, and in this county Lewis, Angell & Cattier.

Last Tuesday I secured Lewis & co. and set them to work. Got a martial band from Marietta and went to work stumping the County. Have spoken every night since. Had a meeting 13 miles out last night, preached there this morning, and returned this p.m. I stir up the people to prepare the way for my officers — am acting as a *recruiting Jno. the Baptist,* and believe we will reap where we are now sowing. I expect to get 25 or 30 men out of this county. We have got the thing started — the hardest, you know, with three names, and many more are anxious ....

I propose making Lewis chaplain, if he does not succeed in securing a lieutenancy. He is a noble man, full of energy, faithful and consistent as a Christian, and a persistent worker. He will secure at least 15 men, has given up his circuit, and thrown himself heart and soul into our cause. Again, his brother is a lieutenant in one of the 6 months' regiments whose time will expire this coming winter. He will have influence with his brother; perhaps we may use it to advantage. He is a very good preacher. I hope this arrangement will meet with your approval.

My "air-castling" for Company K is as follows —

I expect Manchester to secure at least	30
Lewis & his Athens Co. co-workers	30
Krug	5
Barber	2
Hess	3
Heikes	12
Vallendar	10
Trumbull Co. &c	5
	97
And three officers	3
	100
Add for Hatfield	3
	103

You name Dilley, Barnes, Morrow,[63] Chapman,[64] & Donaldson[65] for promotion. Do you mean that they are to be the first promoted? Please state definitely what vacancies you count on them to fill. Please leave Co. K open. I may be compelled to use every position. Is the above the order you desire observed?

I enclose you one of my bills. I shall keep on stumping until I hear from you. Meantime I will get things shaped up, ready for my departure.

Present my warmest congratulations to "Ad" & Bates for their bravery, and my condolence and sympathy to Yeomans and Barnes. Committing you to the care of God, I remain

Ever devotedly, D.H. Moore

Athens, O., Oct. 16, 1863

My Dear Colonel: "We have met the enemy and he is ours!" Ohio [in its gubernatorial election October 13] has buried in eternal condemnation the cause and advocates of treason! Vallandigham, convicted by two courts, damned by the people, "waits and watches over the border," and will wait till summoned by Gabriel's trumpet to meet his final doom. O this victory is grand! It tells how exhaustless is the patriotism of the good state of Ohio! It tells her soldiers they are not forgotten! It tells the President that all her influence and resources are at his command! It tells the haughty South that we still are bent on war! It tells all Europe that we are not divided — that they must not, dare not, interfere! O how glorious! 75,000 majority, home vote! Our little county gave 1780 of it!

How complete the downfall of copperheads! Of the 76 democratic votes polled in this township, only two had Vallandammer's name scratched; and yet, yesterday, those ballots were claimed by over *forty!* They are crawling into their holes and tugging away, trying to pull the holes in after them!

One good thing follows another. I enclose Tod's circular, notifying the people that the draft will occur the 26th inst! I *hope* to have K filled to the minimum by that time. I have fifty reported now, and have not heard from Vallendar. Smith 1, Barber 1, King 1, Chase 11, Lewis 12, Manchester 24 = 50.

I have written a dozen letters which remain unanswered. I have only heard once from you since the battle [of Chickamauga]. I fear my letters have miscarried. I am *exceedingly* anxious to hear from you. I have been "waiting and watching" — so far in vain. Hope to hear to-morrow night. I am stirring up all my recruiters to a final and grand effort. I want K filled by volunteers so that the conscripts may be assigned [to] the old companies. Have you any thing late and definite from Company I?[66]

Ever Faithfully,
D.H. Moore

Athens, O., Oct. 25, 1863

My Dear Colonel: Yours of 14th is just at hand. I will answer it to-day, for in the morning I start for Columbus and Cleveland to make arrangements for organizing Company K ....

I am glad the Governor has asked for worthy sergeants to re-

L. M. Strayer Collection

A September 1863 enlistee, Richard Roessler served two years in Company I
as a private, corporal and sergeant. In this Greenwall & Stringham portrait
a white 4th Corps badge was affixed to his blouse.

cruit for commissions. They will work with a will. Then, too, I am glad the Adjutant will get a "leave." I wish he were here now to take my place. Moses, if it were not that he is detailed for conscripts, could take it. Capt. Vallendar has the majorship on the brain; the trust would *spoil* him. He is a good fellow, though, and whipped a Dutch Copperhead just before [the] election — an act that, in my eyes, covers a multitude of sins. He is a good recruiter, next to Dr. Mc the best. I wish you would appropriate a neat sum for his use out of the regimental fund. His expenses are heavy. If those detailed are ordered to report to Capt. Elmer Otis,[67] Supt. Ohio Volunteer Recruiting, Columbus, it will be better for them about drawing their pay.

It is my intention to break for "Dixie" just as soon as K is organized. I then can leave things in "ship shape" so that our interests will not suffer. With K, I hope to take what conscripts we may realize from the first draft.

Thank God for the January prospect! 35,000 from Ohio! That will fill us sure, veteran companies and all. To recruit between the two drafts is to be the work of the details.

I have just recruited a young officer — a 1st Lt. in 75th O.V.I. until a few months since, when he resigned on account of ill health. He is a fine officer, great energy, brave and experienced. Has soldiered 21 months. I shall make him a sergeant in K, and, perhaps, leave him to recruit. If the Sergt. Major's position, or that of commissary or Q.M. sergt. is vacant, it would be well, perhaps, to wait and give him a trial. Please bear this in mind.

I have 76 now reported for K. Expect the company will be a minimum one. Shall save the conscripts for the old companies. Smith is useful here, more so than any of the recr. detail now here except Vallendar. I will send him on the first of Nov., unless our company goes the week thereafter.

Your horse is now in Cleveland in W.H. Potts' stables *on expense*. If I can't get transportation for him, I will send Smith right down with him ....

I shall act upon your principal and definite recommendations before leaving for Dixie, viz. — Moses to be Capt. E; Barnes 1st Lt. A; King 1st Lt. B; Clark 1st Lt. H. *Between us,* if there is any prospect of a vacancy in B soon I know Moses would rather wait and have that company. That would make a place for Adj. too.

Your splendid description of the battle came safely and has afforded great gratification. I am under promise to read it to-night to the Editor of our paper. How my heart swells with pride at Genl. Wood's testimony as to the conduct of [the] 125th! God bless you all!

Affy, D.H. Moore

Athens, Nov. 8, 1863 Sunday p.m.

My Dear Colonel: The literal compliance with "Special Order No. 29" on my part for the last two or three weeks has been the result of circumstances — I've been too busy, too much perplexed to write oftener.

I left you in my last ... burdened with all the anxieties and perplexities that were then weighing down my spirits. Well, Monday night showed 57 in camp, with four more in Dayton, five in Ashtabula, four in this county, one in Cleveland, two deserters in Washington Co., whom we hope to secure. Smith's two men and Phillips' one man "played out" — deserted or were taken away from them. I ordered Manchester to proceed at once to muster in a detachment so that he could be commissioned first lt. and be able to draw what the men needed, and then to send detachments out to recruit [and] to report back in ten days.

At 12 that night I started for Athens, ordering Lt. Smith to Dayton to "inspect" Hess & Heikes, and then to proceed to Salem in Washington county to enlist, if possible, one of my old company boys. He performed his work well. Heikes at Dayton bids fair to earn his position. Smith succeeded in enlisting Payne[68] of Salem, through whom I expect to realize at least two more. Meantime, I had secured the services of Capt. Williams[69] of an independent co., through whom I expect to make the 20 already recruited in this county up to forty or fifty. Then last evening we went out some eight miles and enlisted two good fellows and struck a "lead" for more.

Lewis, Williams and I will work hard here this week, and Williams and I with the men recruited will go to camp next Monday, to-morrow a week, when I hope we shall be able to organize "K." If so, I shall start with it the latter part of that week *for you.* It depends upon how many are rejected, and how well the different recruiting squads succeed.

Through Mr. Park we heard of a man at Andover, Ashtabula Co., with a squad of 6 men, and a prospect of making it *ten.* He wanted a second lieutenancy for them! I sent a man to see him and offer him the position of "orderly" [sergeant] in K. Have not yet heard the result. And that's the way K stands *exactly*.

Recruiting is now better than ever. The quota for every township of the January draft is published. The Military Committees are beginning to take hold of it. The desire to free the state of the draft is growing. Great efforts will be made. From this on recruiting will improve until the draft actually begins.

It is not ours to find fault with the Government for not drafting at once. We haven't time. If we stopped to do it, somebody else would get the recruits! We have to accept it as it is, and the motto is, "The best fellow first and the Devil take the hindmost!" Regiments will be more or less nearly filled up in proportion as their representatives are more or less active and successful in recruiting.

The Adjutant and his detail arrived in Columbus Monday and left on Wednesday .... I fixed the sergeants up with recruiting appointments, &c., and they are stationed as follows: Chapman, Youngstown; Steadman, Middlefield, Geauga Co.; Dilley, Howland, Trumbull Co.; Barnes, Bloomfield; Evans, Ravenna; Adjutant, Mt. Vernon.

I understand that the conscript details are all about to be ordered back; if they are not, I shall be afraid to get ours ordered back, and for this reason: If the draft occurs and ours is the only regiment that has none but recruiting details here, they will leave us to the last & perhaps then refuse to turn over a single man to any but a conscript detail — as General Mason told me before Moses &c were sent.

Lt. Smith starts to you this week. Vallendar we must retain. He has no money and can't draw his pay. He fancies Phillips — Phillips has some money — hence I leave him to work with Vallendar, relieving Sergt. Seydler of Co. I.

Yours of 30th ult. is at hand. I fear you had the blues the least bit when you wrote. Remember, "Night brings out the stars." I presume "I" has reached you ere this. Also my letters in reference to chaplaincy. Your horse is all right. He is now in Kirk's stables, Columbus. He is a noble fellow ....

I shall do every thing in my power to get off with K the last of next week. My anxiety to join you increases daily, *and my disgust of recruiting is awful!*

Ever Yours,
D.H.M.

Athens, Ohio, Nov. 16, 1863

My Dear Colonel: The last week has been a busy one. We have done every thing we could to get men, and have succeeded very well, *considering*. So far as heard from, we now stand,

Mustered in	47
Unreported	5
Unreported Dayton	7
Unreported Marietta	2
New Recruits Athens Co.	15
Others – musicians	2
	78

Have not heard from recruiting officers in north part of State, where considerable effort will be put forth to avoid the draft; nor from the Andover man with his ten men. Moreover, we have till next Friday, the 20th, to go upon. I think there will be no doubt but the company will be full by that time. I expect to start for "Dixie" the 25th, unless some unforeseen circumstances prevent.

Was over at Marietta last night. Was serenaded, called out for a speech, and several other *horrid* things occurred. I have invited my old company to meet me at an oyster supper at the National House there next Wednesday night. I shall make a great effort to get my old 2nd lt., Jenkins — of whom I have before written you — to take hold and recruit up the old boys and take a lieutenancy in it. The boys are *partial* to "A," that having been their letter. They are a splendid set of fellows, and I hope to get some more of them. I believe there is a vacancy in A — Is there? If not, there is at least some good place for them.

Among the boys of my old Co. is one, my 3rd Sergt., H[enry] Lord, a splendid penman, a fine business man, sober, honest, at-

Opposite: Charles F. Davis, a farmer residing at Napoleon, Ohio, was recruited by Horace Welch August 22, 1863, but saw little, if any, field service with the regiment. He reported sick soon after Company I's arrival in Chattanooga November 16 and convalesced almost a year at Brown Hospital in Louisville, where this portrait was made at J.C. Elrod's gallery. From February 1865 until his discharge that August he was detached as a clerk in the Louisville office of Assistant Commissary Musters, Department of Kentucky. Earlier in the war Davis, a Medina County native, spent 13 months as a trooper in Company I, 2nd Ohio Cavalry.

John Halliday Collection

tentive, and capable. If the post of Commissary Sergt. is open, I would like to recommend him for it. He will do splendidly. He does not want Sergt. Major. Please write me here *particularly* of this.

Genl. Buell will in all probability be Grant's Chf. of Staff.[70] It would be fine to be in his favor — to have his *open ear.* Mrs. Curtis of Marietta, his sister, is now visiting with him in Lawrenceburg, Ind. Her husband is one of our state senators. Her son was my first corporal.[71] He is full of energy, courage, and *snap,* and would make us a splendid *any thing* from 1st Sergt. up to lieutenant. His penmanship is nothing *extra,* altho' *very good.* So far as the active duties of Sergt. Major are concerned, he would be unsurpassed. What can you do for him? Write particularly, *please.*

Of the men recruited for K, 7 were mustered for old companies, one man had 9 less than he reported, 8 were rejected by surgeon, and six were rejected by mustering officer = 30. And so we have had to work it up. I hope to hear from you soon, and to receive the certificate of the chaplain's election. I am the more anxious because you may have some commands for me before I leave.

Ever Faithfully,
D.H. Moore

Athens, Nov. 18, 1863

My Dear Colonel: Your long expected and most welcome letter of 6th came to hand last night, with the Chaplain's appointment. For this kind and prompt acquiescence in my request I am profoundly grateful. It was essential to the successful termination of my recruiting operations. *Between us,* be it always understood, Bro. Lewis does not fill my bill for chaplain in every respect. He lacks the oratorical power and the impressivenenss of manner one so loves to witness; nor is he so good a penman as he ought to be. On the other hand, he is a humble and sincere christian, kind and social, popular with the people, a very good preacher, and an untiring worker. Moreover, he will not intermeddle with business belonging to head-quarters, as some Chaplains *do;* and he will do all that mortals can do to carry out your wishes. He is green in the service, and scarcely knows the first principle of soldiering; yet, I think we will find him one of the most persevering and industrious and useful of chaplains. He is a good recruiter, and I greatly desire to detain him here until the draft. I thought a regimental order — like mine — would fix him. I shall try to have him detailed by the Adjutant General. I also desire to secure the detail of the 1st Lt. K and a private to recruit for us in Athens Co. until the draft. The three would net us, I *think,* thirty men.

The two "I" sergeants I shall bring with "K." Hatfield has been very sick ever since his return & is only now convalescent. Seydler has done good service with Vallendar. Smith left Columbus for you last Thursday. Phillips I would have sent, but for this reason. Vallendar — a strange coincidence, for he is utterly ignorant of your recommendation — has set his head on having Phillips for his 2nd lt.! V. is strapped, can't collect his subsistence bills or draw his pay, & hence is discouraged. P. has some money, which he furnishes freely, and that makes "Old Lager" all right ....

To-night I give my oyster supper in Marietta. Hope to meet with some success among my old boys. Shall post you upon returning. Friday I start to Columbus with my recruits — 16 reported, but expect to have at least 20 or 25. Hope to effect an organization at once, and start the 25th or 26th for you. As to rags and dirt, I fancy our new clothes will be almost a disgrace among veterans! My wife suggests that, as to the dirt, perhaps you are out of soap, & that I had better take an extra supply!

Soberly — I'll thank God when I am done with this recruiting! It's horrible business! it's disgusting, too! Yet it's all right; it is necessary; my success reconciles me to it. Nevertheless, I'm thankful it will soon be over!

I find that it will be impossible for me to work in any of the old boys as an officer of K. If recruiters in other parts do as well for K as we have done, it will be 90 or 100 strong; if not, it will not vary far from a minimum. A run of bad luck *may* delay us again; but I think it scarcely possible.

Don't worry about the "horse question." It's only pleasure for me to serve you. He will be forth coming all right. Once started for Dixie I shall push ahead as rapidly as possible; but my speed, I suppose, will be proportioned to the condition of the roads.

> Ever Faithfully,
> D.H. Moore

Athens, Nov. 19, 1863

Dear Colonel: Yours of 10th is at hand. Am delighted to hear from you, especially in 9 days. Heretofore — that is, since the battle, 12 & 15 d[ays] were required. It is a pleasing and favorable sign.

You need not worry about the horse. It is a pleasure for me to bring him to you.

Welch is green — impudent, *ignorantly,* I hope. You will have to teach him good manners. His Orderly, except Smith, is the best man in the Company. It is strange they are so long getting through; perhaps, when I come to try it, I may understand the reasons better.

Obedient to your wishes, I will endeavor to have Manchester detailed. I also want a private detailed to assist Lewis in this county; also one for Washington Co. That would pretty effectively "work up" the state. Guess I can make it with Genl. Hill.

As to "buying up" men so as to "make a point," none of that thing has been done yet. Not a man has been bought. The Dayton men together will earn one commission, & only one. The men are mine; the officers must settle the "commissioned" between them. If that involves the necessity of Heikes buying out Hess — I shall say, *"Amen, let it be so!"*

Again, some expense incurred — say $15 by me, assisting Williams to recruit, which is a drop in the bucket to my expenditures, & which was incurred while acting for him. He will have to pay. In so far and no further will the thing be allowed, and this scarcely deserves the name of "buying up." Some one must have written you to that effect or you would not have thought of it. I am gratified that your confidence in me is firm — it shall never be *abused.*

Well, I went last night to Marietta. Some forty or fifty of my old company honored me with their presence. We had oysters etc. in abundance, brass band, and a splendid time. Although I was disappointed in securing any recruits immediately, yet I flatter myself that I have the thing "set" for some of them before the draft. I can't make the "riffle" with Lt. Jenkins. It's all right.

To-morrow I go to camp. Started "Jo. Hooker" to-day. One of my recruits rides him to Columbus. I fear the "Northern" recruiters have not done as well as their "Southern" brethren; if so, another week is added to our deplorably prolonged stay. I shall try to get the men up warm and comfortable.

Good "goak" about my "fat purse!"

I will write you as soon as I see the Columbus development.

Ever Yours,
D.H. Moore

Columbus, Ohio, Nov. 27, 1863

My Dear Col: Once more God be thanked! Another great battle, another great victory;[72] and yourself unharmed. Your praise is in the mouth of every one. You have made yourself immortal. God bless you! God bless every member of the 125th! And let all the people say Amen.

Can scarcely repress my congratulations long enough to give you a satisfactory account of our standing here. We still lack in spite of all our exertion some six men to enable us to muster easily. I have sent Capt. Manchester to Cleveland to report to Capt. Vallendar, hoping that both together may be successful in speedily raising the required number. Arrangements have been made to fully arm and equip the Company as soon as mustered. I much fear that we will not get away from here before next Wednesday. I shall be perfectly satisfied under the circumstances with that.

The Dayton matter was settled by the Adjutant General in favor of Lieut Heikes who is now in camp ....

The advance pay question has been decided in our favor. Every thing is in our hands, except the men. I pine for you and for our "Ohio Tigers" like a caged bird. I am flat on my back to-day three fourths sick and I believe it is my repeated disappointment in getting away that causes it. I enclose a complimentary paragraph from to-day's *Commercial*.[73]

Again congratulating you and the Regiment for your great triumph.

> I Remain as ever Devotedly Yours,
> D.H. Moore
> Lt. Col. 125th O.V.I.

Columbus, Nov. 29, 1863

Dear Col, Give me your hand on the recent victory and the glorious record of the 125th and Col. Opdycke. It is beyond description *grand*. Thank God for such men!

Two reasons conspired to prevent my making one of Col. Banning's company last Friday. First, I was "sick a bed," and second, the public announcement of Capt. Baugh's dismissal. I could not bear to see the Capt. smarting under his disgrace. And so you will please present my excuses.

We have been most perplexingly and provokingly delayed in

the organization of these two companies — the 10th hangs fire almost as badly as the 9th did. We still lack some four men. Capt. Manchester will be in Cleveland recruiting until he gets that number. He will get them in two days, I think.[74]

I expect to get off this week. Shall I have the pleasure of your company?

Cordially,
D.H. Moore

Notes to Appendix

1. Clark, p. 57.
2. David H. Moore CSR.
3. David H. Moore, *An Escape That Did Not Set Me Free: A By-Product of Morgan's Raid. A Paper Read Before the Ohio Commandery of the Loyal Legion April 7, 1915* (n.p., n.d.), p. 4.
4. Opdycke diary.
5. David H. Moore CSR.
6. *Ohio Roster*, vol. VIII, p. 443, 446.
7. *To Battle for God and the Right*, p. 223; Opdycke diary, April 5, 1864.
8. Moore to Opdycke, September 22, 1863, Emerson Opdycke Papers, OHS.
9. David H. Moore CSR.
10. Ibid.
11. *To Battle for God and the Right*, p. 228.
12. Moore's postwar positions included president of Wesleyan College for Women in Cincinnati; editor of the *Western Christian Advocate*, also in Cincinnati; and organizer of the University of Denver.
13. First Lieutenant Edward S. Aleshire, formerly of Company A, 87th Ohio.
14. First Lieutenant William A. Gage, formerly of Company H, 87th Ohio.
15. Stewart formerly was a corporal in Company A, 87th Ohio.
16. Horace Welch of Mount Vernon became first lieutenant of Company I, 125th Ohio. He resigned January 27, 1864.
17. Captain Isaac D. Spaulding, Company D, 125th Ohio, resigned April 6, 1863.
18. Sterling Manchester formerly was the 87th Ohio's sergeant major.
19. John P. Hoyt.
20. Captain Anthony Vallendar of Company H, 125th Ohio, formerly served in the Prussian army. He was a lithographer born in the Rhine River city of Coblenz. Anthony Vallendar CSR, RG 94, NARA.
21. Captain Steen B. Parks of Company F, 125th Ohio, formerly commanded Company I, 85th Ohio. He was wounded May 9, 1864 at Rocky Face Ridge, Georgia, and resigned June 20, 1864.
22. Federal forces under Major General Robert H. Milroy were attacked and defeated June 14-15, 1863 by a portion of Lee's army near Winchester, Virginia. Union losses were 95 killed, 348 wounded and 4,000 reported captured or missing. Long, p. 366.
23. Ohio Adjutant General Charles W. Hill. He was succeeded in January 1864 by Benjamin R. Cowen. Reid, p. 814, 964.
24. Albert Opdycke was born in New Jersey in 1788. *To Battle for God and the Right*, p. xxi.
25. Leonard Eckstein Opdycke, Emerson Opdycke's only child, was born September 26, 1858. *To Battle for God and the Right*, p. xxiii.
26. Moore referred to the successful Tullahoma campaign in Middle Tennessee and Union victory at Gettysburg, Pennsylvania.

27. Sergeant James H. Hanson became first sergeant of Company I, 125th Ohio. He was captured January 17, 1864 at Dandridge, Tennessee, and survived months of confinement at Andersonville prison. Clark, p. 211.

28. At the time, Colonel George B. Senter was Camp Cleveland's commandant.

29. Major General John A. Dix, a Buchanan Administration treasury secretary, assumed Department of the East command July 18, 1863. After the war he served one term as governor of New York. Warner, p. 126; *OR*, vol. XXVII, pt. 1, p. 2.

30. A Columbus hotel.

31. Sergeant John A. Bell, Company E, 125th Ohio.

32. Colonel William E. Gilmore commanded some 550 Ohio militiamen in the pursuit of Morgan. *OR*, vol. XXIII, pt. 1, p. 779.

33. Colonel Benjamin P. Runkle, 45th Ohio Mounted Infantry.

34. Joseph L. Pickering, formerly of Company H, 87th Ohio.

35. In the July 19 fight at Buffington Island, Morgan suffered about 820 casualties, including 700 captured. Long, p. 388.

36. Charles Ballard manufactured farm implements in Athens, and later at Springfield, Ohio.

37. Colonel Roy S. Cluke, 8th Kentucky Cavalry (C.S.), was one of Morgan's brigade commanders late in the raid.

38. Lieutenant Colonel Cicero Coleman, 8th Kentucky Cavalry (C.S.).

39. At the time, Brigadier General James M. Shackelford commanded the 1st Brigade, 2nd Division, 23rd Corps.

40. Colonel August V. Kautz, 2nd Ohio Cavalry, commanded his regiment and the 7th Ohio Cavalry during Morgan's raid. The German-born officer was promoted to brigadier general in May 1864. Warner, p. 258.

41. An expanded account of Moore's experiences during Morgan's raid, based on this letter, is found in his 1915 pamphlet *An Escape That Did Not Set Me Free*. After the war Moore and Coleman stayed in touch. When Coleman died at Lexington, Kentucky, in January 1915, Moore spoke at his funeral.

42. Morgan and 364 of his troopers surrendered July 26 in Columbiana County, Ohio.

43. John Brough of Marietta was elected Ohio's governor on October 13, 1863. He soundly defeated Peace Democrat Clement L. Vallandigham.

44. It is unclear whether Moore's "Chronicle" references were made about Ridgley C. Powers, or an unnamed employee of the Warren newspaper.

45. Major General Thomas L. Crittenden of Kentucky commanded the Army of the Cumberland's 21st Corps.

46. Captain Calton C. Baugh, Company E, 125th Ohio, was dismissed July 31, 1863 for incompetency.

47. Henry Barber, formerly a private in Company F, 36th Ohio.

48. Josiah H. Jenkins, formerly second lieutenant of Company A, 87th Ohio.

49. Dennis Hart Mahan, a U.S. Military Academy professor and engineer, wrote several treatises that greatly influenced Mexican War and Civil War leaders. Boatner, p. 501.

50. Brigadier General John S. Mason, formerly colonel of the 4th Ohio Infantry, was assigned to recruiting and muster duty in Ohio in April 1863. Warner, p. 314.

51. Although Edward P. Bates temporarily commanded the 125th Ohio on numerous occasions, and served as a brigade quartermaster, he never mustered at a rank higher than captain. In August 1865 the War Department bestowed upon him the brevet ranks of major, lieutenant colonel and colonel for "gallant and meritorious services during the war." Edward P. Bates CSR.

52. Martin H. Hamblin had resigned January 5, 1862 as captain of Company G, 41st Ohio.

53. Captain Frederick A. Nash was provost marshal of Ohio's 18th District, with headquarters in Cleveland.

54. At the time, Major General David S. Stanley was the Army of the Cumberland's chief of cavalry.

55. A blacksmith by profession, Aquila Coonrod was born in Williams County, Ohio. Aquila Coonrod CSR, RG 94, NARA.

56. James B. Steedman, later promoted to brigadier and major general.

57. Colonel Peter J. Sullivan resigned August 7, 1863. *Ohio Roster,* vol. IV, p. 433.

58. Moore referred to the just-fought battle of Chickamauga, in which the 125th Ohio suffered 105 casualties. *OR,* vol. XXX, pt. 1, p. 176. According to the regimental history, another 100 were "more or less injured" but remained on duty. Clark, p. 131.

59. Company I was mustered September 25. Its original officers were Captain Aquila Coonrod, First Lieutenant Horace Welch and Second Lieutenant Seabury A. Smith. *Ohio Roster,* vol. VIII, p. 443.

60. Private Gustave Seydler, Company I, 125th Ohio.

61. Samuel Heikes became second lieutenant of Company K, 125th Ohio.

62. Sergeant Edward Hatfield, Company I, was wounded at Franklin and reduced to private July 1, 1865. *Ohio Roster,* vol. VIII, p. 445.

63. After his right thigh was shattered at Missionary Ridge, Sergeant John A. Morrow, Company H, was discharged for disability August 6, 1864. Clark, p. 179; *Ohio Roster,* vol. VIII, p. 440.

64. Charles C. Chapman, a saddler, was a resident of Youngstown. Enlisting as a private in Company G, he rose through the ranks to become Company I's captain in July 1865. Charles C. Chapman CSR, RG 94, NARA.

65. Sergeant Henry A. Donaldson, Company E, was promoted to second lieutenant of Company H and first lieutenant of Company G. *Ohio Roster,* vol. VIII, p. 432, 437, 440.

66. Company I did not arrive in Chattanooga until November 16, 1863. Clark, p. 147.

67. Mustering officer Elmer Otis belonged to the 4th U.S. Cavalry.

68. Oren V. Payne, formerly a private in Company A, 87th Ohio, was appointed corporal in the 125th's Company K. Wounded at Resaca, he was promoted to sergeant in June 1864. *Ohio Roster,* vol. VIII, p. 446.

69. Waldern S. Williams was appointed Company K's first lieutenant,

and in February 1865 became captain of Company E.

70. An unfounded rumor. On October 24, 1862, Major General Don Carlos Buell, a native of Washington County, Ohio, was relieved of Army of the Ohio command and succeeded by fellow Ohioan William S. Rosecrans. After waiting 17 months for new orders, Buell mustered out of volunteer service and resigned his regular commission June 1, 1864. Although General Ulysses S. Grant recommended Buell's restoration to duty, he was not recommissioned and soon retired to business pursuits. Warner, p. 52.

71. Corporal William F. Curtis Jr., Company A, 87th Ohio.

72. Moore referred to the battle of Missionary Ridge, November 25, 1863.

73. *Cincinnati Daily Commercial.*

74. Company K finally was mustered December 5, 1863. Its original officers were Captain Sterling Manchester, First Lieutenant Waldern S. Williams and Second Lieutenant Samuel Heikes. *Ohio Roster,* vol. VIII, p. 446.

BIBLIOGRAPHY

Manuscript materials

Opdycke, Emerson. Papers, MSS 554, Archives-Manuscripts Division, Ohio Historical Society, Columbus.
Vallandigham, George B. Letters, Timothy R. Brookes Collection, East Liverpool, Ohio.
Whitesides, Edward G. Diary, *Civil War Times Illustrated* Collection, U.S. Army Military History Institute, Carlisle Barracks, Pennsylvania.

Newspapers

Ashtabula Weekly Telegraph
Cleveland Herald
Ohio State Journal (Columbus)
The Jeffersonian Democrat (Chardon)
Western Reserve Chronicle (Warren)

Published letters, diaries & memoirs

Fellman, Michael, editor, *Memoirs of General W.T. Sherman,* New York: Penguin Books, 2000.
Hazen, William B., *A Narrative of Military Service,* Boston: Ticknor and Company, 1885. Reprinted by Blue Acorn Press, 1993.
Hood, John Bell, *Advance and Retreat,* Edison, N.J.: The Blue and Grey Press, 1985.
Longacre, Glenn V. & Haas, John E., editors, *To Battle for God and the Right: The Civil War Letterbooks of Emerson Opdycke,* Urbana: University of Illinois Press, 2003.
Moore, David H., *An Escape That Did Not Set Me Free: A By-*

Product of Morgan's Raid. A Paper Read Before the Ohio Commandery of the Loyal Legion April 7, 1915, n.p., n.d.

Rice, Ralsa C., *Yankee Tigers: Through the Civil War with the 125th Ohio,* edited by Richard A. Baumgartner & Larry M. Strayer, Huntington, W.Va.: Blue Acorn Press, 1992.

Government records & publications

Compiled Service Records of Volunteer Union Soldiers who Served in Organizations from the State of Ohio. Records of the Adjutant General's Office, 1780-1917, Record Group 94, National Archives and Records Administration, Washington, D.C.

Heitman, Francis B., *Historical Register and Dictionary of the United States Army,* 2 volumes, Washington: Government Printing Office, 1903.

Illinois. *Report of the Adjutant General of the State of Illinois,* 8 volumes, Springfield: Phillips Bros. State Printers, 1900-1902.

Kentucky. *Report of the Adjutant General of the State of Kentucky,* 2 volumes, Frankfort: John H. Harney, Public Printer, 1866-1867.

Ohio. Roster Commission. *Official Roster of the Soldiers of the State of Ohio in the War of the Rebellion, 1861-1866,* 12 volumes, Akron, Cincinnati, Norwalk, 1886-1895.

Records of the Office of the Judge Advocate General – Army, Court-Martial Case Files 1809-1894, Record Group 153, National Archives.

Terrell, W.H.H., *Report of the Adjutant General of the State of Indiana,* 8 volumes, Indianapolis: W.R. Holloway, State Printer, 1865-1868.

United States. Adjutant General's Office. *Official Army Register of the Volunteer Force of the United States Army for the Years 1861, '62, '63, '64, '65,* parts I-VIII, Gaithersburg, Md.: Olde Soldier Books Inc., 1987.

United States War Department. *The War of the Rebellion: A Compilation of the Official Records of the Union and Confederate Armies,* Washington: Government Printing Office, 1880-1901.

Unit histories

Clark, Charles T., *Opdycke Tigers 125th O.V.I.,* Columbus: Spahr & Glenn, 1895.

Hinman, Wilbur F., *The Story of the Sherman Brigade,* Alliance, Ohio: Press of Daily Review, 1897.

Kimberly, Robert L. & Holloway, Ephraim S., *The Forty-First Ohio Veteran Volunteer Infantry in the War of the Rebellion,* Cleveland: W.R. Smellie, 1987. Reprinted by Blue Acorn Press, 1999.

Wilson, Lawrence, *Itinerary of the Seventh Ohio Volunteer Infantry, 1861-1864,* New York: Neale Publishing Co., 1907.

Secondary works

Biographical Directory of the United States Congress 1774-Present, http://bioguide.congress.gov

Boatner, Mark M. III, *The Civil War Dictionary,* New York: David McKay Company, Inc., 1959.

Boyd's Cleveland Directory and Cuyahoga County Business Directory 1863-64, Cleveland: Fairbanks, Benedict & Co., Printers, 1863.

Boynton, Henry V., *The National Military Park Chickamauga-Chattanooga. An Historical Guide,* Cincinnati: The Robert Clarke Company, 1895.

Cozzens, Peter, *This Terrible Sound: The Battle of Chickamauga,* Urbana: University of Illinois Press, 1992.

Dyer, Frederick H., *A Compendium of the War of the Rebellion,* Dayton: Press of Morningside Bookshop, 1979.

Funk & Wagnalls New Encyclopedia, 29 volumes, Chicago: R.R. Donnelley & Sons, 1983.

Howe, Henry, *Historical Collections of Ohio,* Norwalk, Ohio: The Laning Printing Co., 1896.

Hunt, Roger D. & Brown, Jack R., *Brevet Brigadier Generals in Blue,* Gaithersburg, Md.: Olde Soldier Books Inc., 1997.

In Memoriam, Henry Glenville Shaw, California Commandery MOLLUS, circular no. 14, March 15, 1907.

Long, E.B., *The Civil War Day by Day: An Almanac 1861-1865,* Garden City, N.Y.: Doubleday & Company, 1971.

Reid, Whitelaw, *Ohio in the War: Her Statesmen, Generals and Soldiers,* 2 volumes, Cincinnati: The Robert Clarke Company, 1895.

Ryan, Daniel J., *The Civil War Literature of Ohio,* n.p., 1911.

Sword, Wiley, *Embrace An Angry Wind,* New York: Harper Collins, 1992.

The National Cyclopaedia of American Biography, New York: James T. White & Company, 1906.

Walker, Charles M., *History of Athens County, Ohio,* Cincinnati: Robert Clarke & Co., 1869.

Warner, Ezra J., *Generals in Blue: Lives of the Union Commanders,* Baton Rouge: Louisiana State University Press, 1993.

INDEX

Page numbers in boldface indicate photographs or illustrations.

Adairsville, Ga., 206
Adams, Comfort A., 30
Adams Express Company, 113
Alabama troops,
Infantry:
24th Alabama, 120
28th Alabama, 120
Alcorn, James Lusk, 29, 35
Alcorn State University, 35
Aleshire, Edward S., 234, 275
Allatoona Hills, 156
Amberg, Louis, 249, 251
Andersonville prison, Ga., 72, 135, 276
Andover, Ohio, 266, 268
Andrews, Joseph, 60, 73
Annapolis, Md., 18
Antietam, battle of, 33, 239
Appomattox, Va., 212, 223
Army corps (U.S.),
1st Corps, 176
4th Corps, 35, 73, 105, 109, 120, 133, 139, 141, 143, 144, 150, 154, 158, 166, 169, 170, 176, 177, 179, 180, 181, 183, 184, 188, 189, 190, 195, 196, 201, 208, 212, 224, 225, 227, 264
9th Corps, 133
11th Corps, 105, 120
12th Corps, 105, 120
14th Corps, 80, 90, 102, 121, 144, 150, 154
15th Corps, 120
20th Corps (1863), 72, 103
20th Corps (1864), 156, 177, 182
21st Corps, 75, 77, 79, 102, 276

23rd Corps, 133, 144, 146, 150, 159, 184, 188, 196, 276
Reserve Corps, 73
Army of Kentucky, 71, 73
Army of Northern Virginia, 75, 102, 223
Army of Tennessee, 34, 103, 139, 160, 211
Army of the Cumberland, 16, 21, 28, 63, 66, 72, 73, 75, 76, 79, 81, 83, 90, 94, 105, 116, 138, 141, 142, 158, 161, 259, 276, 277
Army of the Ohio, 278
Army of the Potomac, 102, 105, 139, 176, 208, 238, 239
Army of the Tennessee, 105, 120, 178
Ashland, Ohio, 240
Ashtabula, Ohio, 162, 164, 181, 235, 261, 266
Ashtabula County, Ohio, 14, 59, 73, 235, 239, 241, 249, 266
Athens, Ala., 188, 196
Athens, Ohio, 21, 138, 231, 234, 239, 241, 242, 243, 244, 245, 247, 251, 252, 253, 255, 257, 260, 266, 268, 270, 271, 276
Athens County, Ohio, 231, 241, 262, 268, 271
Athens County Militia, 242
Atlanta, Ga., 17, 22, 27, 140, 160, 161, 173, 177, 178, 180, 183, 184, 185, 186, 187, 196, 204, 206, 234
Atlanta campaign, 28, 142, 176
Augusta (Georgia) Railroad, 180
Austinburg, Ohio, 239, 249

Baird, Absalom, 60, 67, 73
Bald Hill (Atlanta), 161
Ballard, Charles, 243, 276
Ballard, Clisby, 152, 159
Banning, Henry B., 18, 20, 33, 53, 72, 231, 251, 273
Barber, Edwin L., 202, 211
Barber, Henry, 249, 261, 262, 263, 276
Barnes, Albert, 95, 104, 262
Barnes, Rollin D., 30, 141, 158, 262, 265, 267
"Barney" (Opdycke's horse), 17, 33, 98, 120, 182
Bascom, Howard, 175
Bates, Edward P., 68, **69,** 70, 74, 103, 108, 111, 117, 126, 136, 137, 144, 173, 186, **187,** 194, 225, 255, 262, 277
Baugh, Calton C., 53, 72, 247, 273, 276
Beatty, Samuel, 214, **215**
Beckwith, George, 129, 138
Beech Grove, Tenn., 80
Beeman, Richard, 54, 72
Belgian muskets, 21
Bell, Henry A., 158
Bell, John A., 241, 260, 276
Benjamin, Judah, 223
Bennison, William H., 60, 73
Bentonville, N.C., 223
Bestow, M.P., 86
Bladensburg, Ohio, 151
Blain's Crossroads, Tenn., 125, 128, 130, 232
Blake, John W., 202, 211
Bloomfield, Ohio, 30, 267
Blue Springs, Tenn., 212, 213, 216
Blyston, David K., **192,** 193, 194, 197
Boner, John, 129
Boyle, Jeremiah T., 40, 71
Bradley, Luther P., **189,** 190, 196, 214, **215**
Brady, Mathew, 207
Bradyville, Tenn., 77
Bragg, Braxton, 34, 75, 78, 80, 81, 82, 90, 103, 110, 111, 139, 239
Brentwood, Tenn., 59, 73

Bridgeport, Ala., 113
Brock Field (Chickamauga), 103
Brookfield, Ohio, 175
Brough, John, 102, 106, 246, 247, 276
Brown, Isaac, 169
Brown, Isaiah, 218, **219**
Brown, Thomas, 218, **219**
Browne, Charles F., 138
Bruff, Joseph, 106, **107,** 108, 120, 146, 149, 159, 166, 181, 216, 221, 223, 228
Buchanan, James, 276
Buckeye Hotel (Columbus), 241
Buckner, Allen, 145, 158
Buckner, Simon B., 120
Buell, Don Carlos, 61, 270, 278
Buell, George P., 102
Buffington Island, Ohio, 102, 243, 276
Bull's Gap, Tenn., 212, 213, 216, 220
Bullitt, William A., 145, 158
Bunnel, Reuben, 108
Burch, Cassius C., 129
Burnham, Thomas M., 167, 174, 175, 181, 186
Burnside, Ambrose E., 117, 118, 121
Bush, Daniel K., 20
Bushnell, Seth A., 12
Buzzard's Roost, Ga., 147, 148, 187

Cadwallader & Tappen (photographers), 232
Cairo, Ill., 228
Calhoun, Ga., 155
California National Guard, 27
Callahan, Mathias, 129
Calvin, Luther S., 158
Camp Chase, Ohio, 35, 228, 235, 238, 249, 252
Camp Cleveland, Ohio, 15, 17, 18, 20, 21, 49, 60, 79, 224, 276
Camp Dennison, Ohio, 232, 258, 259, 260
Camp Harker, Tenn., 200, 213, 214, 216, 218, 223, 224
Camp Irwin, Texas, 225
Camp Opdycke, Ky., 39, 40, 41
Camp Wickliffe, Ky., 13

Camp Wood, Ohio, 12
Canon, John, 133
Carey, Jesse H., 218, **219**
Carr, Joel, 218, **219**
Carter, Abner B., 120
Carter House (Franklin, Tenn.), 193, 196
Cato, Marcus Porcius, 25
"Cato" (Henry Glenville), 25, 27, 62, 66, 81, 184, 195, 197, 209, 222, 224
Catoosa Springs, Ga., 141, 142, 144
Central District of Texas, 35, 227
"Ceylon" (Ridgley C. Powers), 27, 28, 41, 47, 60, 61, 63, 68, 72, 79, 82, 83, 85, 108, 117, 119, 120, 121, 127, 130, 135, 152, 155, 161, 162, 175, 225
Champion (steamer), 227
Chancellorsville, battle of, 139
Chapman, Charles C., 262, 267, 277
Chapman, George W., **43**
Chardon, Ohio, 30, 33, 159
Charleston, Tenn., 152
Chattahoochee River, 160, 169, 175, 205
Chattanooga, Tenn., 17, 27, 28, 65, 75, 86, 89, 90, 93, 94, 101, 103, 105, 106, 108, 109, 113, 114, 117, 118, 120, 121, 125, 127, 140, 141, 162, 182, 184, 185, 193, 196, 204, 216, 223, 256, 268, 277
Chattanooga Daily Rebel, 81
Chattanooga National Cemetery, 158, 174, 181
Cheatham, Benjamin F., 181, 182
Cheatham Hill, 160
Cheney, Walter, 138
Cheshire, Ohio, 245
Chickamauga, battle of, 22, 23, 27, 28, 34, 72, 75, 87, 89-101, 103, 104, 105, 109, 114, 116, 117, 127, 151, 159, 184, 185, 186, 193, 204, 206, 263, 277
Chillicothe, Ohio, 241, 249
Cincinnati, Ohio, 21, 39, 102, 216, 239, 275
Cincinnati Commercial, 273, 278
Cincinnati Enquirer, 51, 56
Clara Poe (transport), 40, 41, 71

Clark, Charles T., 24, 25, 27, 34, 95, 102, 129, **131**, 211, 265
Cleburne, Patrick R., 181, 195, 197
Clemens, Jeremiah, 211
Cleveland, Ohio, 12, 15, 17, 20, 25, 34, 93, 140, 141, 196, 211, 224, 235, 236, 238, 240, 250, 261, 263, 265, 266, 273, 274, 277
Cleveland, Tenn., 141, 144, 146, 173
Cleveland Herald, 21, 25, 27, 29, 30, 33, 51, 52, 53, 105, 120, 164, 184, 197, 203, 209, 224, 225, 228
Cleveland Morning Leader, 145
Cleveland Plain Dealer, 138
Clinch Mountain, Tenn., 125
Cluke, Roy S., 244, 245, 276
Coal Run, Ohio, 235
Coates, Cassius, 175, 182
Coblenz, Germany, 275
Coleman, Cicero, 244, 245, 276
Collins, Freeman, 162, 181
Columbia, Tenn., 49, 53, 61, 188
Columbiana County, Ohio, 102, 120, 276
Columbus, Ohio, 13, 17, 20, 34, 35, 106, 131, 140, 158, 228, 231, 232, 235, 236, 237, 238, 241, 246, 249, 250, 252, 255, 257, 258, 259, 260, 263, 265, 267, 271, 272, 273, 276
Conrad, Joseph, 196
Converse, Julius O., 146, 155, 159, 165, 175, 177
Cook, Gilbert L., 218, **219**
Coolville, Ohio, 243
Coonrod, Aquila, 259, 260, 277
Cooper, Daniel, **55**
Copperheads, 28, 34, 63, 72, 102, 126, 235, 263, 265
Corinth, Miss., battle of, 33, 260
Corinth, Miss., siege of, 59
Couch, A.J., 129
Coulter, Samuel L., 188, 196
Covert, William A., 57, **58**, 59, 73
Cowen, Benjamin R., 35, 275
Cox, Jacob D., 24, 35, 195
Cranston, James, 218, **219**
Crawfish Springs, Ga., 27, 103
Creighton, William R., 120
Crittenden, Thomas L., 77, 80, 90,

102, 247, 276
Crowell, George W., 34, 196
Cumberland Gap, 127, 216
Cumberland Mountains, 84, 185
Cumberland River, 41, 44, 45, 46
Curtis, Leonard, 145, 158
Curtis, William F. Jr., 278
Cuyahoga County, Ohio, 241

Dallas, Ga., 154, 156, 159
Dalton, Ga., 139, 144, 146, 148, 150
Damascusville, Ohio, 120
Dandridge, Tenn., 22, 125, 127, 128, 130, 133, 135, 138, 139, 187, 193, 204, 206, 276
Dartmouth College, 28
Davis, Charles F., 268, **269**
Davis, George P., 218, **219**
Davis, Jefferson, 81, 82, 214, 220, 223
Davis, Jefferson C., 49, 71, 72, 121
Dayton, Ohio, 239, 249, 261, 266, 268, 272, 273
Defiance, Ohio, 260, 261
Defiance County, Ohio, 158, 241
Delaney, Michael, 182
Delaware County, Ohio, 241
Department of Kentucky, 268
Department of West Virginia, 25
Department of the Cumberland, 74, 214
Department of the East, 276
Department of the Ohio, 121
Dilley, Alson C., 30, 68, 70, 143, **145,** 152, 159, 167, 173, 181, 186, 262, 267
Dilley, Jonathan, 60, 73
Dilley, Lewis S., 145, 150, 152, 159
Dillon, Eli H., 54, 72
District of East Tennessee, 223
District of Western Kentucky, 71
Dix, John A., 239, 276
Donaldson, Henry A., 262, 277
Dover, Tenn., 41, 42, 44, 71
Duck River, 53, 188
Durham Station, N.C., 212
Dyer's Field (Chickamauga), 103

Eaglesport, Ohio, 245

East Tennessee & Virginia Railroad, 216
Elk River, 80
Elliott, Washington L., 201, 202, 214, **215**
Elrod, J.C. (photographer), 268
Emancipation Proclamation, 21
Enfield rifle-muskets, 121, 132
Erie County, Ohio, 241
Etowah River, 156
Evans, Eliza, 174
Evans, Ephraim P., 23, **174,** 182, 267
Ezra Church (Atlanta), 161

Farmington, Ohio, 175
Fawkes, Guy, 188, 196
Federal Knapsack, The, 51, 52, 53
Felton, C.C., 253
Fifishire, Scotland, 196
Fishel, Warren H., 218, **219**
Fitch, Orasmus, 218, **219**
Fitch, Riley, 133
Fitch, William R., 218, **219**
Fobes, F.J., 129
Ford's Theater, 223
Forrest, Nathan B., 45, 50, 57, 71, 73, 82, 161, 181, 188
Fort Creighton, Tenn., 120, 185
Fort Donelson, Tenn., 41, 42, 45, 46
Fort Granger, Tenn., 61, 62, 67, 73, 188, 195
Fort Henry, Tenn., 44
Fort Opdycke, Ga., 177
Fort Sanders, Tenn., 121
Fort Wood, Tenn., 110, 120, 185
Fox, Perrin V., 52, 72
Franklin, battle of, 28, 68, 187, 188, 195, 203
Franklin, Tenn., 23, 25, 27, 28, 47, 49, 50, 51, 52, 53, 54, 56, 57, 59, 60, 61, 64, 66, 67, 72, 73, 75, 76, 82, 118, 184, 186, 189, 190, 191, 193, 196, 197, 206, 214, 277
Fredericksburg, battle of, 139
French Broad River, 128, 130
Fulton County, Ohio, 241

Gage, William A., 235, 242, 249, 251, 257, 275

Galt House (Louisville), 71
Gambier, Ohio, 34, 54
Garfield, James A., 90, 97, 98, 103
Garrard, Israel, 129, 136, 138
Gartner, J.P., 133
Geauga County, Ohio, 14, 18, 29, 30,
   241, 261, 267
Gettysburg, battle of, 102, 139, 176,
   239, 275
Giddings, Harvey, 218, **219**
Gilbert, Charles C., 40, 41, 53, 60,
   61, 67, 71, 75, 77
Gilbert, Lafayette, **65**
Gillis, John, 218, **219**
Gilmore, Emory, 218, **219**
Gilmore, James R., 223
Gilmore, William E., 242, 276
Glenville, Henry, 25, **26,** 27, 51-52,
   105, 184, 196, 197, 224
Goetz, John, **100**
Governor's Guards, 259
Graham, Henry, 129
Grand Army of the Republic, 24
Granger, Gordon, 57, 59, 66, 73, 75,
   79, 108, **109,** 118, 139, 186
Grant, Ulysses S., 24, 74, 102, 105,
   139, 185, 213, 221, 239, 270, 278
Greene, Ohio, 175
Greenford, Ohio, 196
Greenville, Tenn., 121
Greenwall & Stringham (photo-
   graphers), 256, 264
Griffin, Henry C., 202, 211
Gulf of Mexico, 225

"H," 29, 30, 113
Hager, Asa, 175, 182
Hall, Asahel B., 40, 71
Hamblin, Martin H., 255, 257, 261,
   277
Hamden, Ohio, 241
Hanson, James H., 237, 238, 249,
   276
Harding, Abner C., 42, 45, 71
Harker, Charles G., 75, 79, 94, 95,
   103, 105, 108, **109,** 110, 116, 118,
   128, 137, 139, 144, 146, 147, 148,
   150, 152, 153, 156, 159, 167, **168,**
   170, 172, 181, 182, 187, 216

Harman, Ohio, 249, 261
Harmon, Heman R., 56, 72
Harpers Ferry, Va., 18
Harpeth River, 49, 52, 57, 59, 60, 61,
   67, 73, 186
Harrison's Landing, Tenn., 117
Hartford, Ohio, 68, 74, 120
Harvard College, 28
Harwood, Avery, 175
Hascall, Milo S., 32
Hatch, Nathan B., 162, 181
Hatfield, Edward, 261, 262, 271, 277
Hazen, William B., 12, 13, 14, 32,
   110, 120
Heikes, Samuel, 261, 262, 266, 272,
   273, 277, 278
Henry, Wallace J., 218, **219**
Henry County, Ohio, 241
Henry repeating rifles, 34, 52
Hess, Joseph, 239, 249, 251, 254,
   258, 261, 262, 266, 272
Hill, Archibald, 41, 71
Hill, Charles W., 68, 236, 239, 272,
   275
Hillsboro, Ohio, 241
Hillsboro, Tenn., 77, 79, 80, 81, 84,
   85, 86, 88, 232
Hiwassee River, 118
Hockingport, Ohio, 243
Holston River, 118, 133
Hood, John Bell, 103, 138, 159, 160,
   161, 176, 184, 186, 188, 190, 194,
   195, 197, 201, 202, 203, 208, 211,
   214
Hooker, Joseph, 109, 110, 114, 116,
   120, 143, 150, 152, 159
Howard, Oliver O., 118, 139, 141,
   142, 146, 147
Howland, Ohio, 30, 173, 186, 267
Hoyt, John P., 235, 275
"Hugo," 29, 30, 113
Hugo, Victor, 29
Huntsville, Ala., 28, 35, 201, 202,
   203, 208, 209, 210, 211, 212
Huron County, Ohio, 72

Illinois troops,
   Artillery:
   Battery C, 2nd Illinois Light, 71

Infantry:
22nd Illinois, 106
27th Illinois, 106, 145, 172, 182
36th Illinois, 194, 225
42nd Illinois, 106, 153
44th Illinois, 194, 211, 225
51st Illinois, 106, 153, 196
73rd Illinois, 194, 220
74th Illinois, 194, 196
78th Illinois, 60, 73
79th Illinois, 106, 116, 120, 145, 158
83rd Illinois, 41, 42, 44, 45, 71
88th Illinois, 194, 196, 202, 211
100th Illinois, 102
Indiana troops,
Infantry:
17th Indiana, 32
40th Indiana, 211, 225
47th Indiana, 32
57th Indiana, 225
58th Indiana, 102
Iuka, Miss., battle of, 33

Jack, David, 60, 73
Jackson, T.J. "Stonewall," 18
Jacob Strader (steamer), 40, 71
Jaquess, James F., 220, 223
Jeffersonian Democrat, 30, 159
Jeffersonville, Ind., 115
Jenkins, Josiah H., 249, 254, 257, 268, 272, 276
Jewell, Jacob, **96**
Johnson, Bushrod R., 103
Johnson, W.T., 240
Johnson's Island, Ohio, 18
Johnston, Joseph E., 139, 141, 146, 148, 154, 155, 156, 160, 161, 165, 166, 170, 212
Jomini, A.H., 12
Jones, Silas N., 54, 72
Jonesboro, Ga., 22, 178, 180, 204, 206

Kautz, August V., 245, 276
Kennesaw Mountain, Ga., 22, 30, 93, 145, 157, 160, 161, 162, 165, 171, 173, 175, 182, 187, 204, 216, 227
Kentucky troops (C.S.),

Artillery:
Cobb's Battery, 120
Cavalry:
8th Kentucky, 276
Kentucky troops (U.S.),
Cavalry:
6th Kentucky, 63, 73
Infantry:
3rd Kentucky, 41, 71, 75, 79-80, 103, 106, 120, 145, 158, 172
Kenyon College, 34
Kershaw, Joseph B., 103
Keys, Joseph H., **151**
Kimball, Nathan, 214, **215**
King, John W., 218, **219**
King, Martin V.B., **95,** 103
Kingston, Ga., 149
Kinsman, Ohio, 48, 72, 102, 174, 186
Knox County, Ohio, 18, 72, 115, 193, 235, 241, 247
Knoxville, Tenn., 113, 117, 118, 121, 125, 127, 128, 130, 133, 135, 208, 216, 218, 223

"Lady Buckner," 110-111, 120
La Grange, Oscar H., 135, 138
Lake County, Ohio, 14
Lake Erie, 203
Lane, John Q., 196
Lawrenceburg, Ind., 270
Lee & Gordon's Mill, Ga., 90, 91, 186
Lee, Robert E., 89, 90, 117, 119, 121, 195, 197, 212, 213, 216, 220, 221, 239, 275
Lee, Stephen D., 195, 197
Leeson's Gallery, 96
LeRoy, F.L. (photographer), 19, 187
Les misérables, 29
Lewis, John W., **163,** 181, 261, 262, 263, 266, 270
Lewis, N.E. (photographer), 93
Lexington, Ky., 276
Liberty, Ohio, 162
Licking County, Ohio, 96
Likens, Richard P., 129, 138
Lilienthal Gallery, 55, 100, 135
Lincoln, Abraham, 13, 21, 25, 34, 72, 102, 155, 221, 223
Ling, Conrad, 129, 138

Little Harpeth River, 52, 190
Little Kennesaw Mountain, **166**
Liverpool, England, 25
Long, Si, 136
Longstreet, James, 75, 89, 93, 95, 103, 118, 121, 125, 126, 127, 130, 133, 135, 137, 138, 187
Lookout Mountain, 105, 108, 109, 110, 111, 113, 114, 120, 185, 205
Lorain County, Ohio, 241
Lord, Henry, 268
Loring, William W., 195, 197
Los Angeles, Calif., 29
Loudon, Tenn., 139
Louisville, Ky., 12, 34, 39, 40, 41, 44, 46, 71, 127, 140, 216, 260, 268
Loutzenhisar, Thomas, 218, **219**
Lovejoy's Station, Ga., 179, 180, 206
Loveland, Ohio, 241, 255
Lucas County, Ohio, 241
Lynchburg, Va., 208
Lynnville, Tenn., 188

MacArthur, Arthur Jr., 194, 196
MacArthur, Douglas, 196
Macon & Western Railroad, 177
Mahan, Dennis Hart, 276
Mahan, John D., 129
Mahoning County, Ohio, 14, 15, 103, 120, 196, 226, 241
Manchester, Sterling, 162, **163,** 164, 165, 181, 235, 237, 239, 247, 249, 250, 251, 254, 258, 261, 262, 263, 266, 272, 273, 274, 275, 278
Manchester, Tenn., 80
Mansfield, Ohio, 100
Marietta, Ga., 160, 162, 164, 165, 175
Marietta, Ohio, 191, 232, 261, 268, 270, 271, 272, 276
Marietta National Cemetery, 171
Marion County, Ohio, 241
Martin, William T., 138
Mason, John S., 252, 254, 267, 276
Mathews, Albert, 218, 219
McArthur, Ohio, 235, 249
McCook, Alexander McD., 91, 93, 103
McCook, Daniel Jr., 66, 73

McHenry, Henry, 54, 72, 135, 141, 241, 252, 254, 265
McIlvain, Alexander, 145, 158
McKee, Samuel, 40, 41, 71
McPherson & Oliver (photographers), 65
Meacham, Norris, 218, **219**
Meade, George G., 102, 238, 239
Mealy, E.W. (photographer), 115
Mecca, Ohio, 28, 102, 175
Medina County, Ohio, 268
Meigs County, Ohio, 243
Melick, Jefferson, **134,** 135
Merrill, William E., 62, 73
Mexican War, 87, 138, 211, 276
Michigan troops,
Cavalry:
2nd Michigan, 129, 132
Infantry:
1st Michigan Engineers & Mechanics, 52, 72
13th Michigan, 102
19th Michigan, 73
Middlefield, Ohio, 29, 267
Military Division of the Mississippi, 105, 139
Miller, Charles, **256**
Miller, Oscar O., **179**
Miller, William, 108
Milroy, Robert H., 236, 275
Missionary Ridge, Tenn., 17, 22, 28, 30, 68, 105, 106, 108, 109, 110, 111, 112, 114, 115, 116, 117, 120, 127, 185, 187, 204, 205, 206, 256, 277, 278
Missouri troops (U.S.),
Infantry:
15th Missouri, 196
Mitchell, Charles D., 136, 138
Mitchell, James, 252
Moffatt, Alexander, 86
Moore, David H., 21, 23, 106, 125, 129, 132, 137, 145, 146, 148, 149, 153, 158, 159, 164, 165, 166, 168, 181, 182, 231, 232, **233,** 234, 275, 276, 277, 278
Moore, Julia, 138
Morgan, John Hunt, 81, 102, 161, 181, 232, 240, 241, 242, 243, 244,

245, 246, 247, 276
Morris, James B., 93
Morristown, Tenn., 216
Morrow, John A., 262, 277
Morrow County, Ohio, 54, 72, 241
Morse, Apollos P., 218, **219**
Morse, A.S. (photographer), 214
Moses, Elmer, 111, 113, 120, 165, 167, 174, 175, 181, 182, 186, 247, 248, 249, 250, 255, 259, 261, 265, 267
Mossman, John C., **48**, 218, **219**
Mount Vernon, Ohio, 18, 34, 54, 55, 72, 135, 193, 267, 275
Mud Creek, Ga., 159, 206
Murfreesboro, Tenn., 11, 21, 34, 39, 41, 68, 72, 75, 77, 79, 80, 118, 216
Muskingum County, Ohio, 171
Muskingum River, 245

Nancy's Creek, Ga., 17, 175
Napier, William, 12
Napoleon, Ohio, 268
Nash, Frederick A., 255, 277
Nashville & Chattanooga Railroad, 61, 86
Nashville, battle of, 28, 201, 203
Nashville, Tenn., 22, 40, 44, 45, 46, 47, 49, 50, 52, 61, 66, 72, 76, 106, 115, 127, 140, 141, 142, 184, 186, 193, 194, 195, 197, 200, 203, 204, 206, 213, 214, 216, 223, 224
Natchez, Miss., 35
National Hotel (Columbus, Ohio), 228
National House (Marietta, Ohio), 268
National Tribune Scrap Book, 24
Naylor, John C., 57, 73
Needs, James A., **99**
Nelson, William "Bull," 13, 39, 71
New Hope Church, Ga., 159, 204
New Market, Tenn., 216
New Orleans, La., 55, 65, 96, 99, 100, 115, 135, 225, 227, 228
New York, N.Y., 207, 225
Newton, John, 139, 166, 170, **176,** 182
Nickerson, William, 54, 72

Norton, Augustus, 136, 138
Noxubee County, Miss., 28

Oakwood Cemetery (Warren, Ohio), 18
Ohio River, 44, 45, 102, 242, 243, 245
"Ohio Tigers," 22, 23, 24, 31, 135, 186, 273
Ohio troops,
Artillery:
6th Ohio Battery, 80
18th Ohio Battery, 73, 104
Cavalry:
2nd Ohio, 245, 268, 276
7th Ohio, 129, 132, 136, 138, 276
9th Ohio, 259
12th Ohio, 260
Infantry:
4th Ohio, 33, 277
7th Ohio, 18, 120, 185
14th Ohio, 260
19th Ohio, 68, 179, 211
26th Ohio, 102, 225
36th Ohio, 248, 276
41st Ohio, 11, 12, 13, 14, 15, 17, 21, 97, 104, 175, 277
45th Ohio, 181, 276
48th Ohio, 260
64th Ohio, 75, 80, 103, 106, 120, 145, 147, 158, 196
65th Ohio, 75, 79, 80, 106, 120, 153, 188, 196
75th Ohio, 265
84th Ohio, 25, 30
85th Ohio, 18, 20, 24, 34, 275
87th Ohio, 18, 20, 33, 231, 232, 235, 240, 242, 258, 275, 276, 277, 278
93rd Ohio, 128, 129, 132, 135, 136
98th Ohio, 40, 41, 71
103rd Ohio, 145, 150, 159
105th Ohio, 15, 152, 159
113th Ohio, 71
121st Ohio, 40, 41, 71, 72
124th Ohio, 39, 49, 71, 79
125th Ohio, 14, 15, 17, 18, 20, 21, 22, 23, 24, 25, 27, 28, 29, 30, 31, 34, 35, 39, 40, 41, 45, 47, 49, 51,

52, 53, 54, 56, 59, 60, 61, 63, 64, 68, 70, 72, 75, 79, 80, 83, 90, 91, 97, 98, 101, 103, 105, 106, 114, 120, 125, 127, 129, 131, 132, 136, 139, 140, 141, 143, 144, 146, 147, 148, 149, 155, 160, 162, 164, 166, 167, 168, 169, 170, 172, 173, 175, 177, 178, 180, 184, 186, 187, 193, 194, 200, 201, 203, 204, 205, 208, 209, 212, 216, 218, 224, 225, 227, 228, 231, 232, 234, 235, 240, 244, 253, 259, 260, 273, 275, 276, 277
171st Ohio, 18
196th Ohio, 35
Oldroyd, L.K. (photographer), 193
Oostanaula River, 150, 155-156
Opdycke, Albert, 275
Opdycke, Emerson, 11, 12, 13, 14, 15, **16,** 17, 18, 20, 21, 22, 23, 24, 25, 27, 28, 29, 30, 31, 32, 33, 34, 35, 49, 51, 53, 54, 57, 64, 68, 71, 72, 75, 76, 78, 79, 83, 85, 87, 89, 91, 94, 95, 97, 98, 103, 104, 105, 106, 108, 111, 116, 117, 120, 125, 133, 135, 137, 139, 141, 144, 146, 147, 148, 150, 152, 153, 155, 159, 166, 167, 168, 169, 170, 175, 176, 179, 180, 184, 186, 187, 190, 193, 195, 196, 197, **200,** 201, 203, 204, 205, **207,** 208, 210, 214, **215,** 218, 224, 225, 227, 229, 231, 232, 234, 273, 275
Opdycke, Leonard E., 32, 275
Opdycke, Lucy, 24, 32
Opdycke, John, 22
"Opdycke Tigers," 23, 106, 169
Orchard Knob, 120
Orrville, Ohio, 43
Otis, Elmer, 265, 277

Pacific Veteran, 27
Paint Rock (transport), 113, 114
Palmer, John M., 79, 102
Panquett, Theophile, 218, **219**
Park, Servetus W., 14, 202, 235, 257, 266
Parks, Steen B., 235, 240, 241, 254, 260, 275
Payne, Darius W., **191**

Payne, Oren V., 266, 277
Peachtree Creek, Ga., 22, 160, 176, 182, 204
Peck, Almon, 218, **219**
Pelham, Tenn., 77, 80, 85, 86
Pemberton, John C., 74
Pennsylvania troops,
 Cavalry:
 9th Pennsylvania, 72
Perry, William, 103
Perryville, Ky., battle of, 33
Petersburg, Ohio, 103
Pettus, Edmund W., 158
Phillips, Nyrum, 78, **79,** 102, 144, **200,** 241, 247, 251, 252, 261, 266, 267, 271
Pickering, Joseph L., 242, 276
Pigeon Hill, Ga., **166**
Pigott, George, 218, **219**
Pine Mountain, Ga., 156
Pittsburgh, Pa., 20, 93, 97, 240, 246
Polk, James K., 53
Pollock, James M., 218, **219**
Pomeroy, Ohio, 242, 243, 245
Port Lavaca, Texas, 225
Port Republic, Va., battle of, 18, 72
Portage County, Ohio, 12, 17, 65, 235, 241
Porter, Benjamin, 172, 182
Portland, Ky., 41
Post of Franklin, Tenn., 53, 59, 60
Potomac River, 25, 239
Potts, W.H., 265
Powers, Milo, 28
Powers, Ridgley C., 28, **29,** 35, 51, 70, 105, 111, 178, 180, 186, **200,** 204, 205, 211, 225, 255, 261, 276
Pulaski, Tenn., 188

Raccoon Mountain, 105
Randolph, Joseph F., 54, 72
Ravenna, Ohio, 18, 162, 174, 182, 235, 267
Reed, Philo E., 42, 71
Reid, William P., 41, 71
Resaca, Ga., 22, 100, 149, 150, 151, 153, 154, 155, 158, 169, 187, 204, 206, 277
Rhine River, 275

Rice, Ralsa C., 24, 25, 186, 218, **219,** 229, 261
Rice, Robert J., 162, 181
Richland County, Ohio, 18, 235, 241
Richmond, Va., 94, 135, 203, 211, 220, 223, 239
Ringgold, Ga., 120, 143, 144
Robertson, Jerome B., 103
Robinson, Gideon A., 218, **219**
Rocky Face Ridge, Ga., 22, 144, 147, 155, 162, 204, 205, 206, 275
Roessler, Richard, **264**
Roper's Knob, Tenn., 67, 68
Rosecrans, William S., 21, 34, 39, 41, 47, 49, 59, 64, 68, 72, 75, 76, 77, 84, 90, 97, 98, 103, 105, 238, 239, 247, 259, 261, 278
Ross County Militia, 242
Rossville, Ga., 97
Rough and Ready, Ga., 178
Ruger, Thomas H., 188, 195, 196
Runkle, Benjamin P., 242, 276
Russell, John, 202, 211
Rutland, Ohio, 242
Ryan, Daniel J., 24

Salem, Ohio, 266
Saltsman, T.F. (photographer), 142
Sample, Jesse, 162, 181
Sandusky, Ohio, 18
San Francisco, Calif., 27
Savannah, Ga., 196, 203
Schenectady, N.Y., 28
Schofield, John M., 188, 195, 196, 197, 214
Scioto County, Ohio, 261
Scott, James B., 129
Scott, James F., **112**
Second Bull Run, battle of, 32
Senter, George B., 238, 247, 276
Sequatchie Valley, Tenn., 86, 88
Sergeant, John W., 34
Seydler, Gustave, 261, 267, 271, 277
Shackelford, James M., 119, 121, 245, 276
Shelter tents, 80, 102
Sheridan, Philip H., 105, **109,** 110, 137, 139, 187, 206, 220, 221, 225, 227

Sherman, William T., 22, 109, 110, 116, 118, 120, 139, 140, 146, 149, 156, 158, 160, 161, 170, 176, 184, 185, 187, 188, 196, 202, 203, 206, 211, 223
"Sherman's Flankers," 177
Shiloh, Tenn., battle of, 11, 13, 14, 59, 175, 260
Simpson, John, 136, 138
Slack, James R., 32
Slocum, Henry W., 180, 182
Smith, George W., 194, 196
Smith, Green Clay, 67, 74
Smith, Seabury A., 78, 102, 117, 129, 132, 133, 136, 138, 241, 247, 251, 252, 254, 257, 260, 261, 263, 265, 266, 267, 271, 272, 277
Smithland, Ky., 41
Snodgrass Hill (Chickamauga), 22, 27
Southington, Ohio, 162
Spaulding, Isaac D., 235, 238, 246, 250, 275
Spencer's Creek, Tenn., 52
Spring Hill, Tenn., 52, 57, 73, 189, 193
Springfield, Ohio, 276
Springfield rifle-muskets, 21, 34, 39, 246, 250
Stanley, David S., 166, 181, 195, 201, 214, **215,** 258, 277
Starnes, James W., 50, 72
Steadman, Hezekiah N., **29**, 30, 105, **200,** 210, 227, 267
Steedman, James B., 260, 277
Steele, Reuben M., 120
Stevenson, Ala., 27, 216
Stewart, Alexander P., 197
Stewart, John V., 235, 275
Stewart, Robert B., 194, 196, 246, 250
Stinger, Daniel A., 165, 181, 255
Stoneman, George, 213, 223
Stones River, battle of, 21, 34, 41, 59, 71
Stones River National Cemetery, 72, 73
Strahl, William, 52
Stratton, Henry G., **179,** 202, 211

Strawberry Plains, Tenn., 130, 133, 135, 137
Strickland, Silas, 196
Sullivan, Peter J., 260, 277
Swinehart, Eli, 144, 158

Taylor, Ezra B., 18
Tennessee River, 67, 80, 105, 113, 184, 185, 214
Tennessee troops (C.S.),
  Cavalry:
  4th Tennessee, 72
Texas troops,
  Cavalry:
  Terry's Texas Rangers (8th Cavalry), 52, 72
Thoman, Freeman, **226,** 227, 228
Thomas, George H., 80, 90, 93, 94, 97, 98, 102, 105, 120, **142,** 143, 184, 186, 202, 203, 208, 222, 224
Thomas, Henry H., 54
Thompson, John, 218, **219**
Thompson, Malcolm, 27
Thompson's Station, Tenn., 63
Thurman, Tenn., 85, 86
Tidd, Mervin, 152, 159
Tod, David, 13, 15, 17, 20, 120, 240, 259, 263
Tracy City, Tenn., 86
Triune, Tenn., 75, 77, 80, 82
Trumbull County, Ohio, 11, 14, 15, 17, 18, 24, 28, 29, 30, 32, 40, 48, 56, 63, 70, 72, 74, 79, 82, 102, 108, 120, 126, 169, 173, 181, 182, 186, 236, 237, 241, 251, 257, 261, 262, 267
Tullahoma, Tenn., 34, 45, 80, 118, 239, 275
Tunnel Hill, Ga., 144, 154, 187
Tupper's Plains, Ohio, 243
Tuscarawas County, Ohio, 238

Union College, 28
United States troops,
  Cavalry:
  4th U.S., 277
University of Denver, 275
University of Michigan, 28
U.S. Christian Commission, 66, 74

U.S. Customs Bureau, 27
U.S. Military Academy (West Point), 12, 59, 73, 276
U.S. Senate, 29
U.S. Signal Corps, 68, 143
U.S. Topographical Engineers, 62

Vallandigham, Clement L., 34, 83, 102, 263, 276
Vallandigham, George B., 34, 102
Vallendar, Anthony, 227, 235, **237,** 240, 241, 250, 254, 261, 262, 263, 265, 267, 271, 273, 275
Van Derveer, Ferdinand, 214, **215**
Van Dorn, Earl, 25, 57, 62, 63, 66, 73, 82
Venango (river steamer), 44, 71
Veteran Reserve Corps, 72
Veteran Volunteer Act, 138
Vicksburg, Miss., 70, 74, 78, 102, 239
"Victor" (Hezekiah N. Steadman), 29, 30, 31, 101, 104, 105, 149, 157, 169, 177, 180, 207
Vinton County, Ohio, 235

"W," 29, 30, 45, 51, 57, 77, 89, 90, 143, 158, 186, 196
Wagner, George D., 89, 102, 145, 158, **189,** 190, 195, 196
Ward, Artemus, 132, 138
Warden, Nathan, 133
Warren *Constitution,* 56, 73
Warren, Jones K., 218, **219**
Warren, Ohio, 11, 13, 14, 17, 18, 19, 20, 24, 25, 30, 32, 71, 73, 102, 139, 140, 179, 187, 211, 225, 235, 236, 276
Washington, D.C., 207, 223, 225
Washington County, Ohio, 235, 266, 272, 278
Waterman, George, 175
Waterman, Sylvester, 175, 182
Waters, C.R., 235
Watkins, Louis D., 63, 73
Wayne County, Ohio, 43, 72, 241
Weimer, William G., **171**
Weiss, John (photographer), 99
Welch, Horace, 235, 236, **237,** 238,

246, 249, 251, 254, 257, 258, 268, 272, 275, 277
Welch, Patrick, 175, 218, **219**
Wesleyan College for Women, 275
West Chickamauga Creek, 75, 89, 90, 116, 205
West Point (U.S. Military Academy), 12
Western Christian Advocate, 275
Western Reserve, 18, 113
Western Reserve Chronicle, 14, 15, 21, 25, 27, 29, 30, 50, 63, 70, 105, 120, 141, 158, 162, 184, 225, 253
Western Reserve Seminary, 28
Wharton, John A., 71
Wheeler, Joseph, 42, 45, 71, 161, 181, 182
Wheeling, Va., 182, 245
Whitesides, Edward G., 20, 23, 91, **92**, 93, 97, 103, 247, 248, 249, 250, 251, 255
Whitmer, Henry, **115**
Wilder, John T., 73
Williams, Waldern S., 266, 272, 277, 278
Williams County, Ohio, 32, 277
Williamsfield, Ohio, 59

Williamson County, Tenn., 52, 190
Winchester, Va., 236, 275
Windham, Ohio, 17
Winstead Hill (Franklin, Tenn.), 196
Wisconsin troops,
Cavalry:
1st Wisconsin, 135
Infantry:
22nd Wisconsin, 73
24th Wisconsin, 194
Witt, M. (photographer), 106
Wood, George L., 18, **19,** 33, 54, 71, 72, 231
Wood, Thomas J., 23, 77, 79, 80, 86, **87,** 88, 94, 98, 102, 103, 105, 120, 137, 185, 201, 214, **215,** 259, 266
Woods, Rufus, 218, **219**
Woodworth, Edwin C., 218, **219**

Yale University, 28
Yeomans, Albert, 49, 72, 95, 104, 196, 262
Yorktown, Va., siege of, 59
Youngstown, Ohio, 267, 277

Zanesville, Ohio, 71

About the Editor

Wisconsin native Richard A. Baumgartner is a former news-paper journalist whose published works include *Buckeye Blood: Ohio at Gettysburg, Blue Lightning: Wilder's Mounted Infantry Brigade in the Battle of Chickamauga, Kennesaw Mountain June 1864, Echoes of Battle: The Struggle for Chattanooga* and *Echoes of Battle: The Atlanta Campaign.* He is a recipient of the Richard B. Harwell and Alexander C. McClurg awards, a past contributor to *Military History* magazine and was a consultant for the Time-Life Books series *Voices of the Civil War.* A full-time researcher and writer, he lives in Cabell County, West Virginia.